Richard Meyes

Emulation of Bursting Neurons in Neuromorphic Hardware based on Phase-Change Materials

Anchor Academic
Publishing

Meyes, Richard: Emulation of Bursting Neurons in Neuromorphic Hardware based on Phase-Change Materials. Hamburg, Anchor Academic Publishing 2015

Buch-ISBN: 978-3-95489-344-7
PDF-eBook-ISBN: 978-3-95489-844-2
Druck/Herstellung: Anchor Academic Publishing, Hamburg, 2015

Bibliografische Information der Deutschen Nationalbibliothek:
Die Deutsche Nationalbibliothek verzeichnet diese Publikation in der Deutschen Nationalbibliografie; detaillierte bibliografische Daten sind im Internet über http://dnb.d-nb.de abrufbar.

Bibliographical Information of the German National Library:
The German National Library lists this publication in the German National Bibliography. Detailed bibliographic data can be found at: http://dnb.d-nb.de

All rights reserved. This publication may not be reproduced, stored in a retrieval system or transmitted, in any form or by any means, electronic, mechanical, photocopying, recording or otherwise, without the prior permission of the publishers.

Das Werk einschließlich aller seiner Teile ist urheberrechtlich geschützt. Jede Verwertung außerhalb der Grenzen des Urheberrechtsgesetzes ist ohne Zustimmung des Verlages unzulässig und strafbar. Dies gilt insbesondere für Vervielfältigungen, Übersetzungen, Mikroverfilmungen und die Einspeicherung und Bearbeitung in elektronischen Systemen.

Die Wiedergabe von Gebrauchsnamen, Handelsnamen, Warenbezeichnungen usw. in diesem Werk berechtigt auch ohne besondere Kennzeichnung nicht zu der Annahme, dass solche Namen im Sinne der Warenzeichen- und Markenschutz-Gesetzgebung als frei zu betrachten wären und daher von jedermann benutzt werden dürften.

Die Informationen in diesem Werk wurden mit Sorgfalt erarbeitet. Dennoch können Fehler nicht vollständig ausgeschlossen werden und die Diplomica Verlag GmbH, die Autoren oder Übersetzer übernehmen keine juristische Verantwortung oder irgendeine Haftung für evtl. verbliebene fehlerhafte Angaben und deren Folgen.

Alle Rechte vorbehalten

© Anchor Academic Publishing, Imprint der Diplomica Verlag GmbH
Hermannstal 119k, 22119 Hamburg
http://www.diplomica-verlag.de, Hamburg 2015
Printed in Germany

Dedicated to

Y.M. and R.M.

In memory of E.D.

Men ought to know that from nothing else but thence [from the brain] come joys, delights, laughter and sports, and sorrows, griefs, despondency, and lamentations. And by this, in a special manner, we acquire wisdom and knowledge, and see and hear, and know what are foul and what are fair, what are bad and what are good, what are sweet and what unsavory. Some we discriminate by habit, and some we perceive by their utility. By this, we distinguish objects of relish and disrelish, according to the seasons; and the same things do not always please us. And by the same organ we become mad and delirious, and fears and terrors assail us, some by night, and some by day, and dreams and untimely wanderings, and cares that are not suitable, and ignorance of present circumstances, desuetude, and unskilfulness. All these things we endure from the brain, when it is not healthy, but is more hot, more cold, more moist, or more dry than natural, or when it suffers any other preternatural and unusual affliction.

attributed to Hippocrates
Fifth Century B.C.E.

Adapted from *The Genuine Works of Hippocrates*, translated from the Greek by Francis Adams, Robert E. Krueger Publishing Co, Huntington, N.Y. 1972

Contents

1. Introduction   1

2. A Biological Background   3
   2.1. The Neuron 3
        2.1.1. Morphology of a Neuron 4
        2.1.2. Ion channels 4
        2.1.3. The Membrane Potential 6
        2.1.4. The Action Potential 10
        2.1.5. Propagation of the Action Potential 12
   2.2. The Synapse 16
        2.2.1. Synaptic Transmission at Chemical Synapses 17
        2.2.2. Synaptic Integration 20
        2.2.3. Synaptic Plasticity 24
        2.2.4. Spike-Timing-Dependent Plasticity (STDP) 27
   2.3. An Overall View 29

3. Experimental Emulations   31
   3.1. Modeling STP and LTP in a CMOS Spiking Neural Network Chip 32
   3.2. Implementation of STDP based on Phase-Change Material Synapses 36
        3.2.1. Phase-Change Materials 37
        3.2.2. A Phase-Change Cross-Point Structure emulating STDP 40
   3.3. Phase-Change Materials for Artificial Neural Networks 44
   3.4. An Overall View 45

4. Bursting Neurons   47
   4.1. Physiological Mechanisms of Bursting 47
   4.2. Bursts as a Unit of Neuronal Information 54
   4.3. Bursting for Selective Communication 55

Contents

    4.4. Modeling Neuronal Bursting Activity 59
        4.4.1. The Integrate-and-Fire Model 59
        4.4.2. The Resonate-and-Fire Model 61
        4.4.3. The Quadratic Integrate-and-Fire model 61
        4.4.4. The Simple Model of Choice 62
    4.5. An Overall View 64

5. A PCM Bursting Neuron   **65**
    5.1. Voltage-Controlled Relaxation Oscillation in a PCM Device 66
    5.2. The Analogy to Hippocampal Pyramidal Bursting Neurons 70
    5.3. Simulation of a PCM Bursting Neuron 77
    5.4. An Overall View 82

6. An Outlook on the Future   **83**

A. Quantification of the Membrane Potential   **87**

B. Vocabulary   **89**

List of Figures   **I**

List of Tables   **V**

Bibliography   **VII**

Acknowledgement   **XVII**

CHAPTER 1

Introduction

In the history of computing hardware, Moore's law, named after Intel co-founder Gordon E. Moore, describes a long-term trend, whereby the number of transistors that can be placed inexpensively on an integrated circuit doubles approximately every two years [1]. Because the number of transistors is crucial for computing performance, significant performance gains could be achieved simply through complementary metal-oxide-semiconductor (CMOS) transistor downscaling. Although Moore's law, which was mentioned for the first time in 1965, turned out to persist for almost five decades, the nano era poses significant problems to the concept of downscaling [2]. Upon approaching the size of atoms, quantum effects, such as quantum tunneling, pose fundamental barriers to the trend. Furthermore, the conventional computing paradigm based on the Von-Neumann architecture and binary logic becomes increasingly inefficient considering the growing complexity of todays computational tasks. Hence, new computational paradigms and alternative information processing architectures must be explored to extend the capabilities of future information technology beyond digital logic. A fantastic example for such an alternative information processing architecture is the human brain. The brain provides superior computational features such as ultrahigh density of processing units, low energy consumption per computational event, ultrahigh parallelism in computational execution, extremely flexible plasticity of connections between processing units and fault-tolerant computing provided by a huge number of computational entities. Compared to today's programmable computers, biological systems are six to nine orders of magnitude more efficient in complex environments [3]. For instance: simulating five seconds of brain activity takes IBM's state-of-the-art supercomputer Blue Gene a hundred times as long, i.e. 500 s, during which it consumes 1.4 MW of power, whereas the power dissipation in the human central nervous system is of the order of 10 W [4, 5]. Thus, it is not only extremely interesting but in terms of computational progress also highly desirable to understand how

Chapter 1: Introduction

information is processed in the human brain. The conceptual idea developed within the framework of this thesis tries to contribute to this intention. In contrast to most recent research dealing with the simulation and emulation of specific connections between nerve cells [5–12], the work of this thesis focuses on investigating, on a purely conceptional basis, the issue of a possible future emulation of an artificial nerve cell based on the inherent physics of phase-change materials.

After this introduction, chapter two provides the reader with the necessary biological background and gives insight into some physiological key processes and functional principles of the nervous system. At some points in this chapter, detailed explanations of selected mechanisms are deliberately left out in order to keep the reader focussed on the central theme of this thesis. Chapter three presents the reader with a selection of recent examples of current research dealing with the emulation of biological functionality. Chapter four describes a specific behaviour of nerve cells which is thought to play an important role in the process of neural information processing and chapter five documents a conceptual idea to emulate this behaviour in an artificial nerve cell based on a phase-change material. Finally, chapter six concludes this thesis and gives an outlook on some future ideas that could be investigated to complement the work of this thesis. Furthermore, keywords that are mentioned for the first time in the text are typed in italic and can be looked up in the vocabulary in appendix B which provides the reader, who might not be deeply familiar with the technical terminology, with the possibility to quickly refresh key definitions which are repeatedly used throughout the whole text. The author hopes to inspire every reader who comes in touch with this field of science for the first time and wishes him to find as much pleasure and excitement in reading this thesis as the author had working on it and writing it down.

CHAPTER 2

A Biological Background

The Human brain is vastly superior to the brain of other animals in its ability to exploit the physical environment in which the controlled organism has to operate. The remarkable complexity of the environment that humans created for themselves since the beginning of their existence depends on the connection of highly sophisticated arrays of sensory receptors to an extremely flexible neural machine - a brain - which provides the possibility to discriminate an enormous variety of events in the environment. The brain organizes the continuous stream of information from these receptors into perceptions which are partly stored in memory for future references. These perceptions are then organized into appropriate behavioral responses. All of this is accomplished by the brain using *nerve cells* that are connected to each other via *synapses*. Even though the nervous system has two classes of cells, *nerve cells (neurons)* and *glial cells (glia)*, which outnumber neurons by a factor of 10 - 50, within the framework of this thesis **only structural and functional properties of neurons are dealt with because neurons are the main signaling units of the nervous system.** [13]

2.1. The Neuron

The Neuron is the basic processing unit of the brain. The human brain contains an extraordinary number of these morphologically simple units (of the order of 10^{11} neurons), each of which has about 10^3 connections to other units. Although classifiable into at least a thousand different types, all neurons share the same basic architecture. Different ways in which neurons with basically similar properties are connected to each other can, nevertheless, lead to quite different characteristics of the resulting entities. The basis for the complexity of human behaviour is formed by the fact that numerous neurons constitute precise anatomical and functional entities rather than by the specialization of individual neurons.

Chapter 2: A Biological Background

In order to appreciate how information in the nervous system is processed it is necessary to begin with with the structural and functional properties of neurons and then to deal with the mechanisms that are responsible for the generation and processing of signals. [13]

2.1.1. Morphology of a Neuron

A typical neuron has four morphologically defined regions, as illustrated in Figure 2.1: (1) the *cell body (soma)*, (2) the *dendrites*, (3) the *axon* and (4) several *presynaptic terminals*. Each region plays a distinct role in the generation of signals and communication between neurons. The soma is the center of *metabolism* of the neuron and has usually two types of extensions: a) several short dendrites and b) one long, tubular axon. Through extensive branching, the dendrites form a *dendritic tree* which functions as the main apparatus for receiving incoming signals from other neurons. In contrast, the axon functions as the main conducting unit that carries signals away from the soma to other neurons. The axon conveys electrical signals in form of *action potentials* (see section 2.1.4) that are initiated at the *axon hillock*, a specialized trigger region at the origin of the axon. The axon itself is partly insulated by *myelin sheathes* that are interrupted at regular intervals by the *nodes of Ranvier*, which enables the fast transport of APs (see section 2.1.5). Near its end, the axon splits in a tree-like fashion into several terminals that form communication sites with other neurons, called synapses (see section 2.2). Presynaptic terminals end mostly at the dendrites of a postsynaptic neuron, however, they may also end at the soma or even at the beginning or the end of the axon of the receiving neuron.

Every neuron's intracellular space is separated from the extracellular space by the *cell membrane* whose *membrane potential* is determined by ion concentrations inside and outside the cell. Changes of the membrane potential can be generated by individual sensory cells in response to smallest stimuli: photoreceptors in the eye respond to a single photon of light; olfactory neurons detect a single molecule of odorant; and hair cells in the inner ear respond to tiny movements of atomic dimensions. **Neuronal signaling in the brain depends on the ability of neurons to respond to such small stimuli by producing rapid changes in the electrical potential difference across their cell membranes. These rapid changes are mediated by *ion channels*, therefore ion channels are important for signaling in the nervous system.** [13]

2.1.2. Ion channels

Ion channels owe their functional importance to three basic properties: (1) they conduct ions, (2) they recognize and select specific ions, (3) they open and close in response to specific electrical, mechanical or chemical signals. Ion channels conduct ions across the cell

2.1. The Neuron

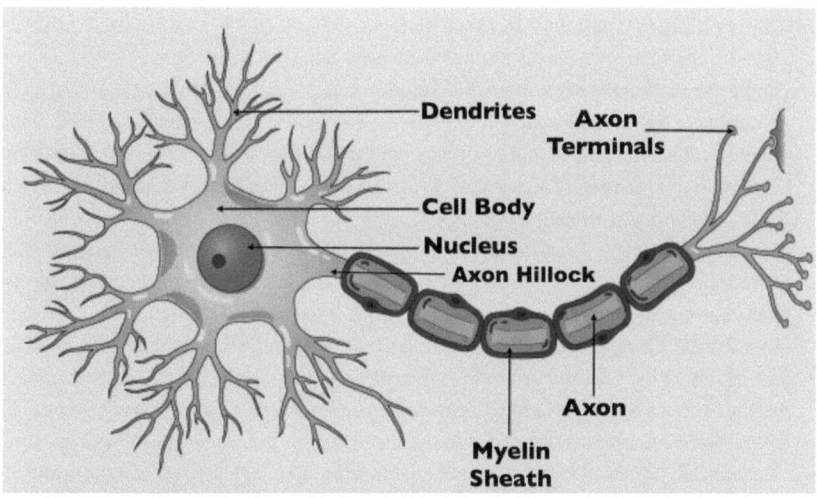

Figure 2.1.: *A typical neuron's morphology. The cell body (soma) is responsible for metabolism processes of the neuron and contains the nucleus, the storehouse of genetic information. It has two types of extensions: (1) several dendrites and (2) the axon. The axon is the signal transmitting element (or the output element) of the neuron and can vary greatly in length. Some can extend up to three meters in the body. Most axons have a relatively thin diameter of about 0.2-20 µm compared to the diameter of the soma (about 50 µm or more). Many axons are partly insulated by myelin sheathes that are interrupted at regular intervals by the nodes of Ranvier which allows the fast transport of APs (see section 2.1.5). Once the signal travelled through the axon it reaches the axon terminals which can connect to other neurons. Such connections, called synapses, mostly appear at the dendrites, the input elements of the neuron and can occur up to a thousand times at a single neuron. [13] [modified from http://insidethemind.synthasite.com]*

membrane between the intracellular and extracellular space at extremely rapid rates: up to 10^8 ions may pass through a single channel per second. Despite the ability to provide high conductance rates, ion channels also provide sophisticated selective mechanisms, i.e. each type of ion channel allows only one or a few types of ions to pass. For instance: the *resting potential* - the membrane potential of a neuron which is at rest, i.e. the neuron shows no

Chapter 2: A Biological Background

activity - is largely determined by ion channels that are selectively permeable to K^+-cations. These K^+-channels are typically 100-fold more permeable to K^+-cations than to Na^+-cations. During *depolarization* (see section 2.1.3), however, Na^+-channels that are 10-20-fold more permeable to Na^+-cations than to K^+-cations are responsible for the value of the membrane potential. The exact understanding of the underlying mechanisms for the selectivity of ion channels is not mandatory for the concept of this thesis, thus, further explanations are left out but can be found elsewhere, e.g. in [13].

The activation or deactivation of many ion channels can be caused by different stimuli, as illustrated in Figure 2.2: (1) *voltage-gated channels* are regulated by changes in voltage, i.e. by changes of the potential across the channel which is determined by the neuron's membrane potential, (2) *ligand-gated channels* are regulated by chemical transmitters, i.e. opening and closing of the channel depends on whether a specific ligand binds at the channel's receptor or not and (3) *mechanically gated channels* are regulated by pressure or stretch. In general, ion channels can enter one of three states under the influence of the above regulating mechanisms: (1) closed and can be activated (resting state), (2) open (active state) and (3) closed and can not be activated (refractory state). The most important task of voltage-gated channels is the generation of APs because the generation and transmission of APs are the basis for encoding neural information in the nervous system. **In order to understand how an AP is generated, it is necessary to begin with a brief dealing of a neuron's membrane potential.** [13]

2.1.3. The Membrane Potential

The electrical signals representing the flow of information in the nervous system are produced by temporary changes in the current flow into and out of the cell, driving the membrane potential - the electrical potential across the cell membrane - away from its resting value. This current flow is controlled by ion channels integrated in the cell membrane, as illustrated in Figure 2.3, whereas these ion channels can be one of two types: (1) *resting channels* and (2) *gated channels*. Resting channels are usually opened and are not significantly influenced by extrinsic factors, i.e. their operational state is not altered by changing e.g. the potential across the membrane. These channels are primarily important in maintaining the resting potential of the neuron, i.e. the electrical potential across the membrane in the absence of signaling. The current that is carried by ion fluxes through resting channels is called *leakage current* and the conductivity of the population of resting channels, which is determined by the amount of ions passing through, is called *leakage conductance*. On the contrary, most gated channels are at rest when the neuron is at rest, i.e. gated channels are closed when the membrane potential is at its resting potential value. In the resting state, the separation of charges across a neuron's cell membrane consists of a thin cloud of ions spread over the inner and outer

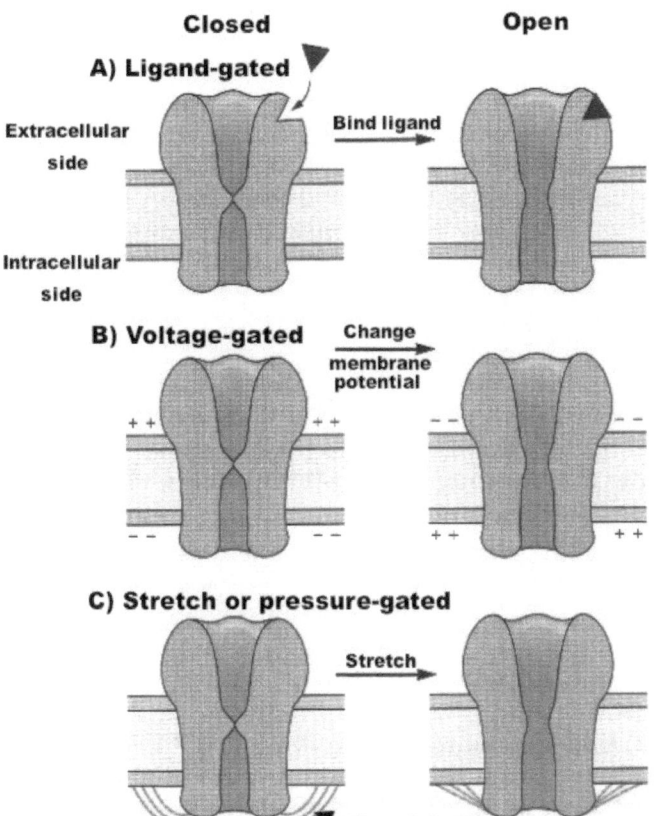

Figure 2.2.: *Several types of stimuli control the opening and closing of ion channels.*
A) Ligand-gated channels open upon binding of a ligand to the channel's receptor.
B) Voltage-gated channels open and close upon changes in the cell membrane potential. The change of the potential causes a conformational change by acting on a component of the channel that has a net charge.
C) Stretch/Pressure-gated channels are activated by stretch or pressure which mechanically forces gating of the channel through the cytoskeleton. [modified from [13]]

Chapter 2: A Biological Background

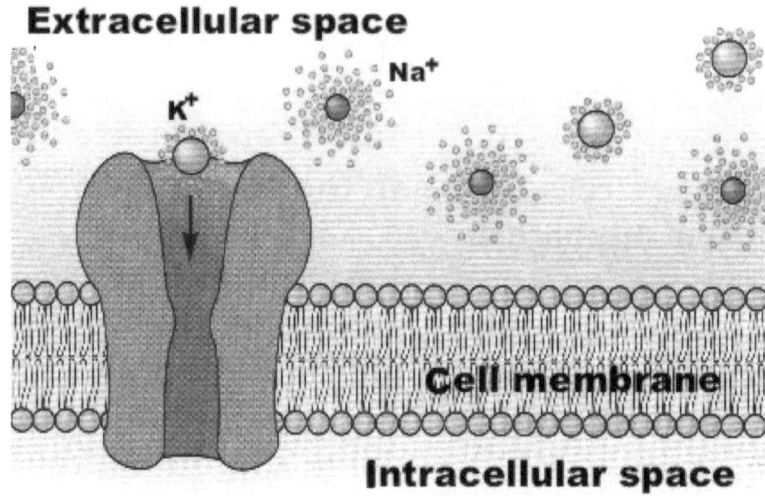

Figure 2.3.: The ionic permeability of the cell membrane is provided by integrated ion channels. These ion channels provide a pathway for hydrated ions to cross the membrane, i.e. ions flow according to their concentration gradient from the extracellular space to the intracellular space and vice versa. [modified from [13]]

surface of the cell membrane. An excess of positive ions on the outside and negative ions on the inside of the cell membrane is maintained because its lipid bilayer blocks diffusion processes of the ions (see Figure 2.4). The resulting charge separation gives rise to different electrical potentials inside and outside the cell defining the membrane potential V_m:

$$V_m = V_{in} - V_{out},$$

where V_{in} and V_{out} are the electrical potentials inside and outside the cell, respectively. Since by convention the potential outside the cell is defined as zero, the resting potential V_r is equal to V_{in} and usually ranges from -60 mV to -70 mV which can be calculated with the Goldman equation (see appendix A).

In order to change the resting potential, electric current carried by both, positive cations (Na^+ and K^+) and negative anions (Cl^- and A^- - organic anions, mostly amino acids and proteins), has to flow into and out of the cell which causes a perturbation of the charge separation and thus, changes the resting potential. A reduction of charge separation leading to a less negative membrane potential is called *depolarization*. An increase in charge separation leading

2.1. The Neuron

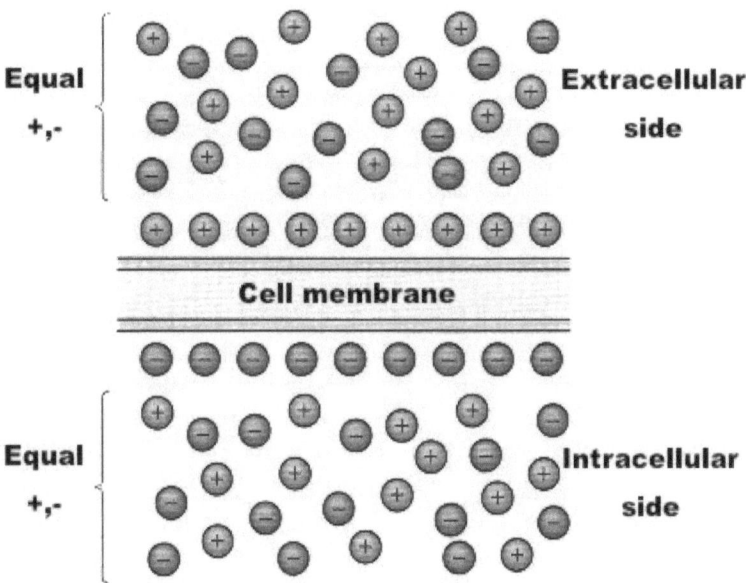

Figure 2.4.: The membrane potential results from a charge separation across the cell membrane. The resting potential is characterized by an excess of positive and negative charges outside and inside the cell, respectively. [modified from [13]]

to a more negative membrane potential is called *hyperpolarization*. In case of perturbation, the membrane potential recovers to its resting potential value thanks to a specific distribution of several resting channels integrated in the cell membrane accompanied by the activity of *ion pumps* that balance the passive flux of ions. The resting channels are either permeable only to potassium (resting channels in glial cells) or permeable to potassium as well as to sodium and chloride (resting channels in nerve cells). The ion pumps prevent the dissipation of ionic gradients by moving ions *against* their net electrochemical gradient. In order to do so, ion pumps need to generate energy which is achieved through hydrolysis of ATP (Adenosine Triphosphate, a multifunctional nucleoside triphosphate used as a coenzyme to transport chemical energy within cells for metabolism) molecules. Thus, the resting potential is not an equilibrium, but rather a *steady state:* the continuous passive influx of Na^+ and efflux of K^+ through resting channels is exactly counterbalanced by the ion pumps. An exception poses the distribution of chloride ions whose movement tends toward equilibrium across the membrane so that there is no net Cl⁻-flux at rest. The exact understanding of the underlying mechanisms for the maintenance of the resting potential, especially the mechanism of ATP hydrolysis which is responsible for the energy extraction of the ion pumps, is not mandatory for the concept of this thesis, thus, further explanations are left out but can be found elsewhere, e.g. in [13]. **When the resting potential is sufficiently perturbed, an action potential is generated, i.e. the balance of ion fluxes that maintains the resting potential is abolished.** [13]

2.1.4. The Action Potential

Depolarization of a neuron's membrane mostly occurs at the dendrites which transport the input signals to the soma (see section 2.1.5). The soma acts as an integrator, spatially and temporally adding up all single input signals from all dendrites. When the membrane potential is depolarized past the *threshold potential*, i.e. the membrane potential rises past a critical value which leads to the activation of voltage-gated ion channels, the balance of ion fluxes in the resting state changes. Voltage-gated Na^+-channels open rapidly in an *all-or-nothing* fashion resulting in an increased membrane permeability to Na^+-ions. The Na^+-influx exceeds the K^+-efflux which leads to a net influx of positive charge causing further depolarization resulting in the activation of additional Na^+-channels which increase the Na^+-permeability even more and so fourth. This regenerative, positive feedback cycle develops explosively, driving the membrane potential toward the Na^+-*equilibrium potential*, i.e. toward the equilibrium which would adjust incase of permanently opened voltage-gated Na^+-channels, of about +55 mV, which can be calculated with the Nernst equation (see appendix A). After the generation of such an action potential (AP), two processes lead to *repolarization* of the membrane potential, i.e. the AP is terminated and the resting potential will be restored: (1) the voltage-gated Na channels gradually close, reducing the Na^+-influx and (2) voltage-gated K^+-channels that were opened during the late stage of depolarization increase the K^+-efflux. The existence of a threshold potential is based on the fact that small depolarizations do not only lead to an increase of Na^+-influx but also to an increase of K^+-efflux which resists the depolarization action of the Na^+-influx up to a certain point. It is important to note that the increase in K^+-permeability during depolarization is much slower compared to the explosive increase in Na^+-permeability because of the slower rate of opening of K^+-channels compared to Na^+-channels. After the AP peak is reached, the delayed K^+-efflux combined with the decreasing Na^+-influx leads to a net efflux of positive charge which continues until the resting potential is restored.

In most neurons, the AP is followed by the *after potential*, a transient hyperpolarization driving the membrane potential toward the K^+-equilibrium potential of about -75 mV. The

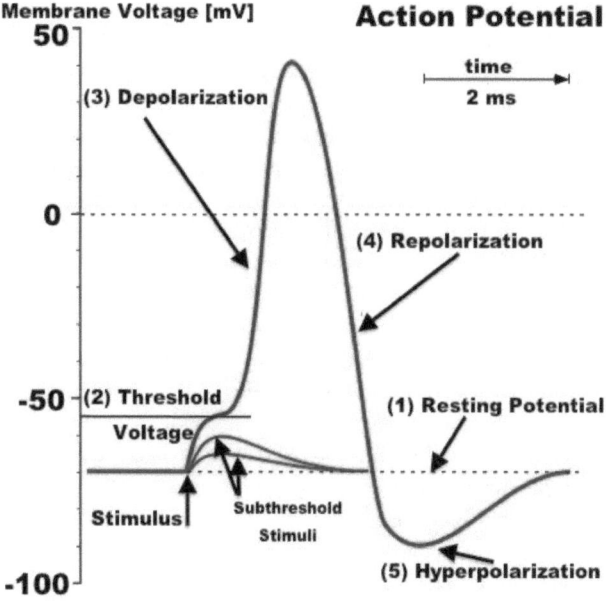

Figure 2.5.: *The change of the membrane potential during the generation of an AP can be divided into five phases: (1) the neuron is at rest, i.e. shows no activity. The resting potential is maintained by the balance of ion fluxes provided by several resting channels and ion pumps; (2) subthreshold stimuli depolarize the membrane and may add up to one single stimulus until the threshold voltage is reached. If no superthreshold stimulus is applied, the resting potential is restored; (3) the membrane potential rises past its threshold value triggering a regenerative, positive feedback cycle of inward Na^+-flux generating the actual AP; (4) the closing of Na^+-channels and delayed opening of voltage-gated K^+-channels drive the membrane potential back to its resting value; (5) the delayed closing of K^+-channels lead to hyperpolarization after which the resting potential is restored. Note that each AP is followed by a period of refractoriness during which the neuron is insensitive to stimuli and can not be excited. [modified from [14]]*

after potential occurs because the K$^+$-channels, which opened during the later phase of the AP, need a few milliseconds to close and are still opened even though the membrane potential has already reached its resting value. Simultaneously, the AP is also followed by a brief period of refractoriness (*refractory period*), i.e. a period during which it is impossible or exceedingly difficult to excite the neuron, that can be divided into two phases: (1) the *absolute refractory period* immediately follows the AP. During this period the neuron is not at all excitable no matter how great the applied stimulating current is. (2) the *relative refractory period* directly follows the absolute refractory period. During this period it is again possible to excite the neuron but the stimuli must be stronger than those usually required to trigger the neuron, i.e. to rise the membrane potential past the threshold value. Both periods of refractoriness are the result of the residual inactivation of Na$^+$-channels and increased opening of K$^+$-channels. It takes a few milliseconds for the voltage-gated Na$^+$-channels that are responsible for the generation of APs to be closed during which they are insensitive to opening signals, thus, leading to the period of refractoriness. Figure 2.5 illustrates how the membrane potential changes during the generation of an AP which is a so called all-or-nothing event, i.e. the underlying mechanisms for the generation are always the same, thus, every AP of a particular neuron looks the same. A neuron's sole ability to generate APs is not enough to process information in the nervous system. **For communication purposes, the neuron has to transport the AP through its axon to the axon terminals, where it can be transmitted to other neurons.** [13]

2.1.5. Propagation of the Action Potential

In order to communicate with other neurons, a neuron has to transport its informational content, i.e. an AP, to its output apparatus, the axon terminals. Every neuron has three relatively constant, passive electrical properties that affect the electrical signaling: (1) the resting membrane resistance r_m (units of $\Omega \cdot cm$) represents the resistance of ion channels for ions passing through a channel from the extracellular space to the intracellular space and vice versa. The current that is carried by ions passing a channel, i.e. the electrical current passing the resting membrane resistance, is called *ionic membrane current*; (2) the membrane capacitance c_m (units of farads) represents the capacitive characteristic of a neuron's cell membrane to separate charges inside and outside the cell (see Figure 2.4). The current that is carried by ions that change the net charge stored on the membrane is called *capacitive membrane current*; (3) the intracellular axial resistance r_a (units of Ω/cm) represents the resistance for a current that flows along the axon and the dendrites. In electric signaling along dendrites and axons, the non spherical geometry of both compartments causes a subthreshold voltage signal to decrease in amplitude with distance from its site of initiation.

The propagation of electrical signals along dendrites and axons can be best understood with the help of an equivalent electrical circuit (see Figure 2.6) that shows how the geometry of

2.1. The Neuron

the compartments influence the distribution of current flow. If ions flow from the extracellular

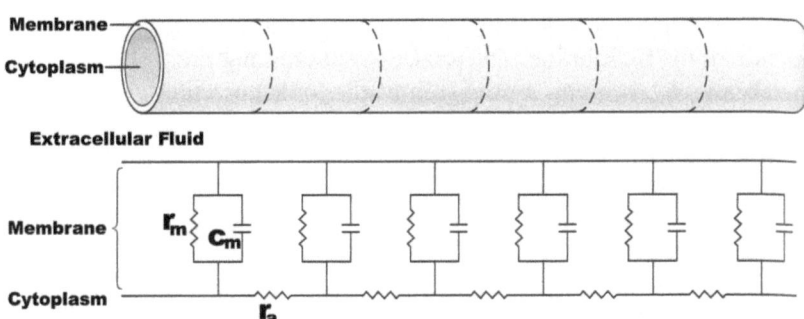

Figure 2.6.: *Equivalent electrical circuit representing a neuronal extension, e.g. a neuron's axon. The extension is divided into unit lengths with an own membrane resistance r_m and a membrane capacitance c_m. The single circuits are connected by resistors r_a, representing the axial resistance of the cytoplasm and a short circuit with negligible resistance representing the extracellular fluid. [modified from [13]]*

fluid into the *cytoplasm* through ion channels, i.e. if a current is injected into the cell and flows through the electrical circuit, which represents a unit length, the current flows out of the cell through several parallel pathways across successive cylinders along the length of the extension. The total resistance r_{tot} for each of these pathways is made of all resistive components in series that the current has to go through on its way into the cell, through the cytoplasm and out of the cell again, i.e.

$$r_{tot} = 2 \cdot r_m + x \cdot r_a,$$

where x is the number of segments along the pathway in the cytoplasm. (Here, for reasons of simplicity, it is assumed that the duration of the current injection is large compared to the time the membrane potential needs to change, i.e. $t \gg \tau_{c_m}$, so that the capacitive current is zero). Because the resistance of the pathways with a greater distance from the site of current injection is bigger, the current I_m decreases along the extension and with it the membrane potential $V_m = I_m \cdot r_m$. Thus, the change of the membrane potential $\Delta V(x)$ depends on the distance from the site of current injection x:

$$\Delta V(x) = \Delta V_0 \cdot e^{-x/\lambda},$$

where λ is the *membrane length constant* and ΔV_0 is the change in membrane potential produced by the current flow at the injection site, i.e. at $x = 0$. The membrane length constant

Chapter 2: A Biological Background

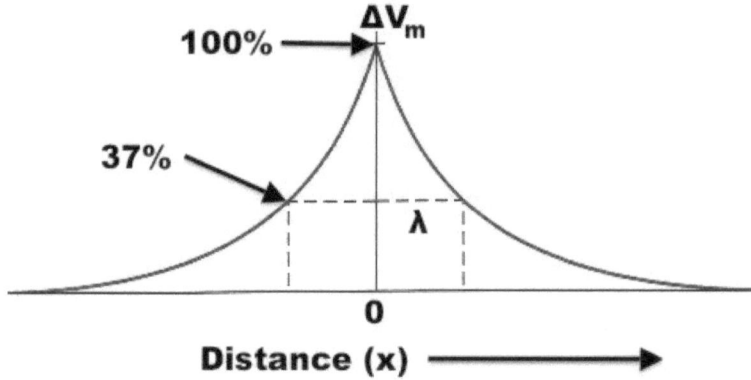

Figure 2.7.: *The change in membrane potential in a passive neuronal extension decays with distance. The distance at which ΔV_m has decayed to 37 % if its initial value is defined by the membrane length constant λ. [modified from [13]]*

is determined by the resistances of the cell,

$$\lambda = \sqrt{(r_m/r_a)},$$

and defines the distance after which the change in membrane potential has decayed to $1/e$, i.e. 37% of its initial value (see Figure 2.7). This means that the better the insulation of the membrane, i.e. the greater r_m, and the better the conducting properties of the cytoplasm, i.e. the lower r_a, the greater the length constant of the extension. The resistances of the cell depend on the cells geometry, more precisely on its diameter, leading to transformed expressions for r_a and r_m:

$$r_a = \rho/\pi a^2,$$

where ρ (in units of $\Omega \cdot$cm) is the specific resistance of a 1 cm^3 cube of cytoplasm and a is the radius of the extension and

$$r_m = R_m/2\pi a,$$

where R_m (in units of $\Omega \cdot$cm^2) is the specific resistance of a unit area of membrane, which leads to an expression for the membrane length constant in terms of the intrinsic (size invariant) properties R_m and ρ:

$$\lambda = \sqrt{\frac{R_m}{\rho} \cdot \frac{a}{2}}.$$

Thus, thicker axons and dendrites have longer length constants than thinner cell extensions and hence, carry electrical signals over longer distances. With the properties of a cell extension

2.1. The Neuron

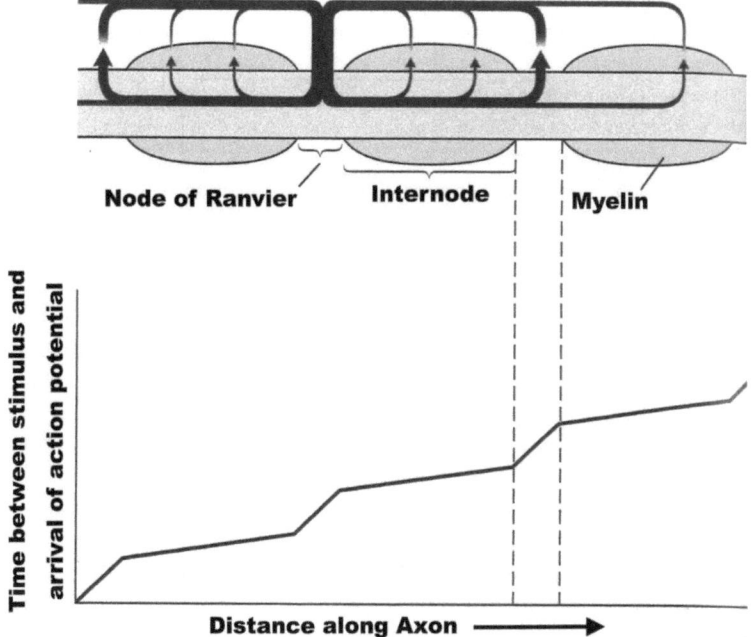

Figure 2.8.: *APs in myelinated fibers are periodically refreshed at the nodes of Ranvier. Capacitive and ionic membrane current densities are much higher at the nodes of Ranvier than in the internodal regions which is represented by the thickness of the arrows. Because of the much higher membrane capacitance at the nonmyelinated nodes, the AP slows down as it approaches each node and thus, appears to jump from node to node. [modified from [13]]*

and their influence on electrical signaling as discussed above, the propagation of an AP through the axon can be understood.

If the membrane at any point of an axon has been depolarized beyond threshold, an AP is generated at this point. The local change in membrane potential spreads down the axon causing the adjacent region to be depolarized past the threshold which leads to the generation of another AP at that adjacent point. The depolarization spreads along the whole axon by *local-circuit current* resulting from the potential difference between the active and inactive regions of the membrane. This current has a great spread in cells with longer length constants leading to a more rapid propagation of APs, whereas there are two ways of increasing the conduction velocity of APs through the axon: (1) an increase in the axons diameter increases the length constant (note the dependence of λ on a) and (2) myelination of the axon, i.e. wrapping of insulating glial cells around the axon, increases the thickness of the axonal membrane and hence, decreases its capacitance. Since the time it takes for a depolarization to spread along the axons is determined by $\tau = r_m \cdot c_m$, partly insulating the axon, i.e. decreasing its membrane capacitance, results in a more rapid propagation of APs. A neuron triggered at the nonmyelinated axon hillock will generate an AP at this point which discharges the capacitance of the myelinated axon ahead of it. The AP is prevented from dying out by the nodes of Ranvier which interrupt the insulation of the axon every 1-2 mm by bare patches of the axon membrane, about 2 μm in length. At these nodes, the AP is refreshed because of the richness of voltage-gated Na^+-channels that generate an intense depolarizing Na^+-inward current in response to the passive spread of depolarization. Figure 2.8 illustrates the propagation of an AP down the axon. Note that the propagation speed of an AP is much faster in the myelinated areas due to the low membrane capacitance. The AP basically jumps from node to node which is called *saltatory conduction*.

In summary, myelination is not only extremely important in terms of conduction speed but it is also favorable from a metabolic standpoint: Because ion channels are integrated only in nonmyelinated parts of the membrane, ionic membrane currents flow only at the nodes in myelinated fibers which means that less energy must be expended by ion pumps in restoring the ion concentration gradients (see section 2.1.3). After an AP travelled through the axon, it reaches the axon terminals which form communication sites with other neurons. **The point at which one neuron communicates with another is called a synapse which is a fundamental element for information processing in the nervous system.** [13]

2.2. The Synapse

Synapses are the connections between the basic units of the nervous systems, the neurons. Synaptic transmission - the transmission of signals from one neuron to another - is the fundamental basis for communication in the nervous system. The human brain contains about 10^{11} neurons, each of which forms about a thousand synaptic connections with other neurons and receives up to a hundred thousand connections, as e.g. the Purkinje cell in in the cerebellum, thus, about 10^{14} or more connections are formed in the human brain. There are more neurons and synapses in one single brain than the several billion stars in our galaxy, fortunately, however, only a few basic mechanisms underlie synaptic transmission at these many connections. A typical synapse consists of a presynaptic axon apposed to a postsynaptic neuron via an axon terminal. Based on the structure of the apposition, synapses

2.2. The Synapse

can be divided into either *electrical synapses*, i.e. the transmission of signals is of electrical nature, or *chemical synapses*, i.e. that the transmission of signals is of chemical nature. At electrical synapses the presynaptic terminal is not completely separated from the postsynaptic neuron so that the current generated by a *presynaptic action potential (PSAP)* flows directly into the postsynaptic neuron through specialized channels called *gap-junction channels* which physically connect the pre- and postsynaptic cytoplasms. In contrast, at chemical synapses the two neurons are physically separated by the *synaptic cleft*. The transmission of signals is provided by the release of *neurotransmitters* upon the arrival of PSAPs at the presynaptic terminals. These transmitters diffuse into the synaptic cleft and bind to postsynaptic receptors, evoking a postsynaptic signal. Several steps that are involved in chemical transmission are the reason for a synaptic delay compared to electrical transmission. Both forms of synaptic transmission can have either inhibitory or excitatory effect on the postsynaptic cell, i.e. that both forms can either facilitate or impede the generation of a postsynaptic AP. Moreover, the strength of both forms can be either enhanced or diminished which is called *synaptic plasticity* and is crucial to memory and other higher brain functions. Table 2.1 summarizes the properties of electrical and chemical synapses. **In the following section, the focus lies on chemical synaptic transmission because, although slower in transmission speed, chemical synapses provide the possibility to amplify signals, unlike electrical synapses, therefore, chemical synaptic transmission is thought to be crucial for emergent phenomena such as memory and learning.** [13]

Type of synapse	Distance between pre- and postsynaptic cell membrane	Cytoplasmic continuity between pre- and postsynaptic cells	Ultrastructural components	Agent of transmission	Synaptic delay	Direction of transmission
Electrical	3.5 nm	Yes	Gap-junction channels	Ion current	Virtually absent	Usually bidirectional
Chemical	20-40 nm	No	Presynaptic vesicles and active zones; postsynaptic receptors	Chemical transmitter	Significant: at least 0.3 ms, usually 1.5 ms or longer	Unidirectional

Table 2.1.: *Properties of electrical and chemical synapses. [taken from [13]]*

2.2.1. Synaptic Transmission at Chemical Synapses

Synaptic transmission at chemical synapses involves several steps, as illustrated in Figure 2.9, starting with the arrival of an AP at the presynaptic terminal. During discharge of a PSAP, voltage-gated Ca^{2+}-channels integrated in the *active zone* of the presynaptic terminal's cell membrane are opened facilitating a Ca^{2+}-inward current. The resulting rise in intracellular Ca^{2+}-concentration causes *synaptic vesicles* to fuse with the presynaptic cell membrane which causes neurotransmitters stored inside the vesicles to be released into the synaptic cleft, a process called *exocytosis*. The neurotransmitter molecules diffuse across the synaptic cleft until they reach their receptors integrated in the postsynaptic cell membrane. Upon arrival at their receptors, the binding of transmitter molecules causes ligand-gated Na^+-channels to be opened facilitating a postsynaptic inward Na^+-current which depolarizes the postsynaptic membrane, thereby generating a *postsynaptic potential (PSP)*. After binding to their receptors, the transmitter molecules must be removed from the synaptic cleft in order to terminate synaptic transmission. In case of no transmitter removal, the synapse would become refractory, i.e. no new presynaptic signals would be transmitted due to receptor desensitization resulting from continuous exposure to transmitter molecules. The understanding of the mechanisms underlying the removal of neurotransmitter from the synaptic cleft is not mandatory for the concept of this thesis, thus, further explanations are left out but can be found elsewhere, e.g. in [13].

The several steps of chemical synaptic transmission are responsible for the synaptic delay which does not occur at electrical synapses. However, the lack of speed at chemical synapses compared to electrical synapses is compensated by the highly important property of *amplification*: A single synaptic vesicle contains several thousand molecules of neurotransmitter, although typically only two molecules of transmitter are required to open a postsynaptic ion channel. Consequently, just a single synaptic vesicle can open thousands of ion channels in the postsynaptic cell. In this way even weak electrical currents generated by small presynaptic nerve terminals of chemical synapses can have much greater impact on the postsynaptic neuron as it would be the case at electrical synapses. After exocytosis, the presynaptic terminal membrane is slightly enlarged, precisely about the size of all vesicle membranes that fused with the terminal membrane, and the number of vesicles inside the cell is decreased. In order to prevent this trend, synaptic vesicle membranes added to the terminal membrane are recycled generating new synaptic vesicles. This recycling process is called *endocytosis* and has not yet been completely understood [13]. Figure 2.10 illustrates the cycling of synaptic vesicles at nerve terminals which involves several distinct steps: (1) free vesicles must be targeted to the active zone and then (2) dock at the active zone after which they (3) become primed in order to undergo exocytosis, i.e. (4) they fuse with the terminal membrane and release the contained neurotransmitter. (5) At last, the fused vesicles' membranes are taken up to the endosome in the interior of the cell by endocytosis where they are regenerated completing the recycling process. A variety of proteins are involved in the recycling process but the exact understanding of the underlying mechanisms for the processes of exocytosis and endocytosis is not mandatory for the concepts of this thesis, thus, further explanations are left out but can be found elsewhere, e.g. in [13].

2.2. The Synapse

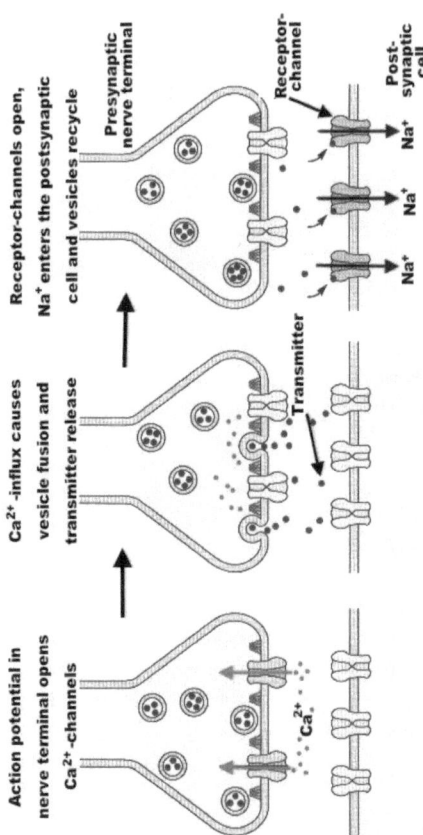

Figure 2.9.: *Chemical synaptic transmission involves several steps: An AP arriving at the presynaptic terminal causes voltage-gated Ca^{2+}-channels at the active zone to be opened which facilitates a Ca^{2+}-influx. The increased concentration in Ca^{2+}-ions inside the cell leads to the process of exocytosis: synaptic vesicles containing neurotransmitters fuse with cell membrane at the active zone causing the transmitter molecules to be released into the synaptic cleft. The transmitter molecules diffuse across the cleft and bind to specific receptors on the postsynaptic cell membrane causing ligand-gated ion channels to open (or close), thereby changing the postsynaptic membrane potential. [modified from [13]]*

Chapter 2: A Biological Background

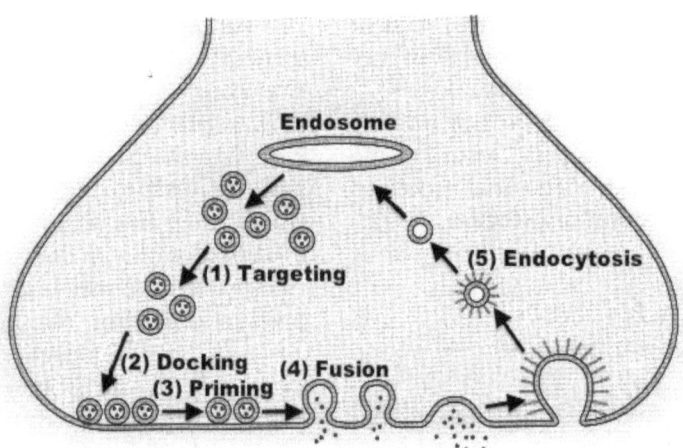

Figure 2.10.: The cycling of synaptic vesicles at nerve terminals involves several distinct steps: Free vesicles are in a first step targeted to the active zone then dock at this active zone in a second step. The docked vesicles become primed in the third step so they can undergo exocytosis. In response to a rise in Ca^{2+}-concentration the vesicles can fuse with the terminal membrane in the fourth step to release the neurotransmitter. After transmitter release, in the fifth step the fused vesicle membrane is taken up into the interior of the cell by endocytosis. The endocytosed vesicles fuse with the endosome (an internal membrane compartment) which regenerates the vesicles completing the recycling process. [modified from [13]]

Postsynaptic neurons usually receive input from about a thousand synaptic connections, all of which can affect the neuron in a different way or with different efficacy. **The neuron has to process all this input simultaneously in order to produce a subsequent reaction, i.e. the neuron has to decide if an AP is generated or not.** [13]

2.2.2. Synaptic Integration

The synaptic input of neurons in the brain, i.e. the *postsynaptic currents (PSPs)* evoked by PSAPs that facilitate synaptic transmission, can affect the postsynaptic neuron in two ways: (1) synaptic input can be excitatory, i.e. an *excitatory postsynaptic current (EPSC)* is generated in the postsynaptic cell. This current is typically carried by Na^+-ions flowing inside the cell through ligand-gated ion channels that are opened after binding neurotransmitter molecules

2.2. The Synapse

at their specific receptors. This inward current depolarizes the postsynaptic membrane in the subthreshold regime generating an *excitatory postsynaptic potential (EPSP)*; (2) synaptic input can be inhibitory, i.e. an *inhibitory postsynaptic current (IPSC)* is generated in the postsynaptic cell. This current is typically carried by Cl$^-$-ions flowing inside the cell through ligand-gated ion channels. This inward current hyperpolarizes the postsynaptic membrane generating an *inhibitory postsynaptic potential (IPSP)*. Consequently, the effect of a synaptic potential is not determined by the type of transmitter released from the presynaptic neuron but rather by the type of ion channels in the postsynaptic neuron gated by these transmitters. Nevertheless, some transmitters act predominantly on receptors that are of one or the other type. For instance: in the vertebrate brain, glutamate (glutamic acid, $C_5H_9NO_4$) as a neurotransmitter typically acts on receptors that produce excitation, whereas GABA (γ-Aminobutyric acid, $C_4H_9NO_2$) and glycine (NH_2CO_2COOH) typically act on receptors that produce inhibition, however, an exception is found in the vertebrate retina and many others can be found in invertebrates [13].

Interestingly, a synapses morphology seems to be correlated to its functionality: two common morphological types of synaptic connections can be found in the brain, Gray type I and type II synapses (named after E.G. Gray). Type I synapses are often glutamatergic, i.e. the presynaptic terminal releases glutamate as neurotransmitter, and therefore excitatory, whereas type II synapses are often GABA-ergic, i.e. the presynaptic terminal releases GABA as neurotransmitter, and therefore inhibitory. Furthermore, excitatory and inhibitory synapses have favored docking sites at postsynaptic cells: Gray type I synapses, often excitatory, preferably form communication sites at postsynaptic somas or dendrites, either at the dendritic shaft itself or at a dendritic spine, a fine specialized input zone of the dendrite, whereas Gray type II synapses, often inhibitory, preferably form communication sites at the postsynaptic neuron's axon. Therefore *axosomatic synapses* and *axodendritic synapses* are often excitatory, whereas *axoaxonic synapses* are often inhibitory. Figure 2.11 and Figure 2.12 illustrate the morphological difference between Gray type I and type II synapses which are also summarized in Table 2.2 and their preferred docking sites at a postsynaptic neuron. A neuron receives synaptic input from about a thousand connections, each of which may be different from the other in terms of whether the input is excitatory or inhibitory, in terms of input strength and in terms of input frequency. In order to decide whether a postsynaptic AP is generated in response to the various competing inputs from all synaptic connections, the inputs are integrated by the postsynaptic neuron which is called *neuronal integration*, a decision-making process which the neurophysiologist and noble laureate Charles Sherrington regarded as the brain's most fundamental operation. Neuronal integration involves the summation of synaptic potentials that passively spread to the trigger zone, the axon hillock, whereas this summation is spatial as well as temporal. *Spatial summation* is the integrative process of summing up synaptic inputs at different communication sites of the postsynaptic neuron,

Chapter 2: A Biological Background

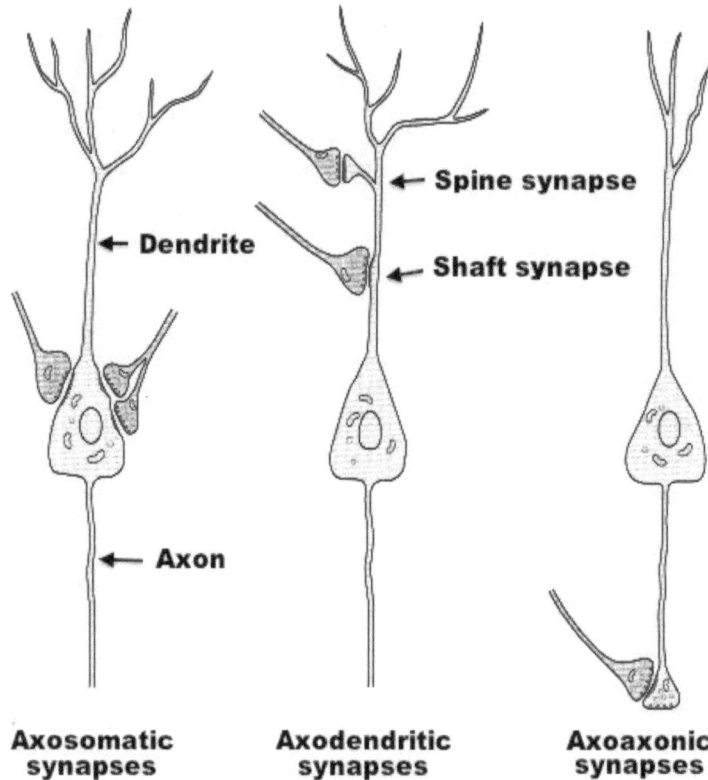

Figure 2.11.: *Synaptic contact can occur at the soma, the dendrites or the axon of postsynaptic neurons. The names of the various kinds of synapses identify the docking regions of the presynaptic terminal at the postsynaptic neuron. Note that axodendritic synapses can either dock at the main shaft of the dendrite or at a specialized input zone, the spine. [modified from [13]]*

whereas *temporal summation* is the integrative process of summing up consecutive synaptic potentials at the same communication site. If the integrative processes of both, temporal and spatial summation result in a superthreshold depolarization of the postsynaptic neuron, an AP would be generated. A remarkable feature of synapses is their ability to undergo functional

2.2. The Synapse

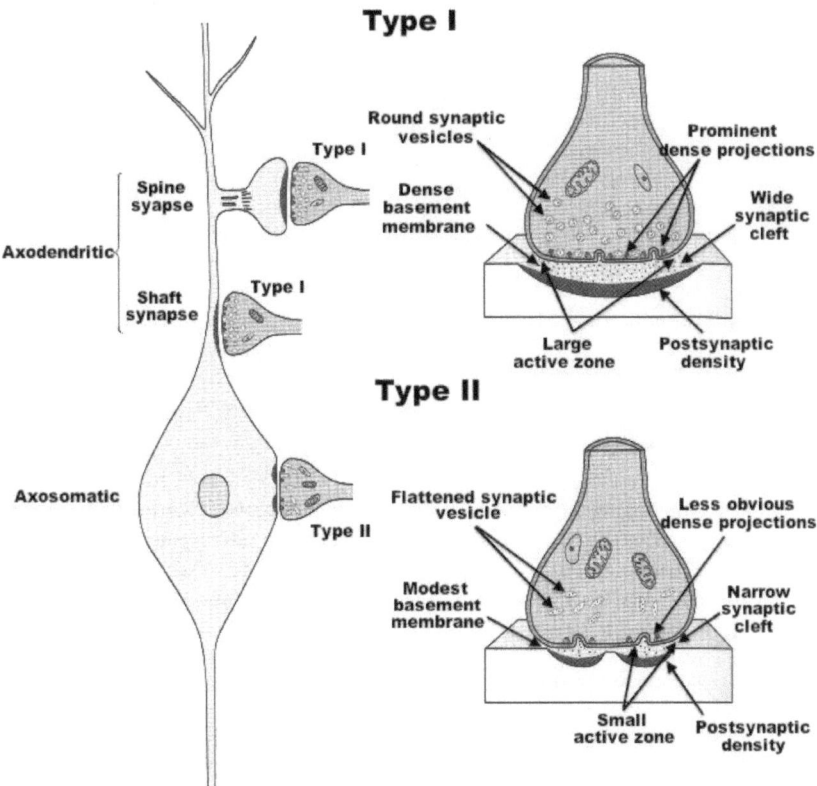

Figure 2.12.: The two most common morphologic types of synapses in the brain are Gray type I and type II. Type I synapses are usually glutamatergic and therefore excitatory whereas type II synapses are usually GABA-ergic and therefore inhibitory. Both types have differences in width of the synaptic cleft, total area of the active zone, prominence of presynaptic densities, shape of vesicles, and presence of a dense basement membrane. Type I synapses commonly contact dendritic spines and sometimes the dendritic shaft, whereas type II synapses commonly contact the postsynaptic soma. [modified from [13]]

Chapter 2: A Biological Background

Type of synapse	width of synaptic cleft	area of active zone	density of vesicle release sites	vesicle shape	density of basement membrane
Gray type I	about 30 nm	$1-2\,\mu m^2$	dense regions on presynaptic membrane; predominant	tends to assume a characteristic round shape	amorphous dense basement-membrane material appears in synaptic cleft
Gray type II	about 20 nm	$<1\,\mu m^2$	less obviously distinct; clustered	tends to assume a rather oval shape	little or no basement membrane in synaptic cleft

Table 2.2.: Summary of the morphological differences between Gray type I and type II synapses.

and structural changes depending on their history in a neural network. **This ability is called synaptic plasticity and is crucial to memory, learning and other higher brain functions.** [13]

2.2.3. Synaptic Plasticity

"What fires together, wires together." - attributed to C.J. Shatz [15].

In 1949, the psychologist Donald Olding Hebb introduced the *Hebbian theory* which explains the adaption of neurons in the brain during the process of learning [16]. The theory states that an increase in *synaptic weight*, i.e. an increase in synaptic efficacy in response to PSAPs, arises from repeated and persistent stimulation of the postsynaptic neuron through a presynaptic neuron. Hebb's theory is often summarized as "What fires together, wires together" and attempts to explain associative learning, an ability of the nervous system to adjust the connection strength of neural pathways, or more precisely the mechanisms by which simultaneous activation of neurons leads to a pronounced increase of synaptic strength between those neurons. This method of learning, named after its originator Donald O. Hebb, is called *Hebbian learning*. Synapses that are affected by Hebbian learning undergo functional and structural changes, called synaptic plasticity, which can be temporary or permanent. A temporary change in synaptic efficacy lasting a few seconds or less is categorized as *short-term plasticity (STP)*. Short-term synaptic enhancement is often differentiated into three categories depending on their timescales: (1) *short-term facilitation (STF)*, also called *pulsed pair facilitation (PPF)* usually lasts for tens of milliseconds while (2) *short-term augmentation (STA)* acts on the timescale of seconds and (3) *short-term potentiation (STP - not to be confused with short-term plasticity)* which has a time course of tens of seconds up to several minutes [17]. Each form of temporary enhancement of synaptic strength is an exclusively presynaptic mechanism and results from an increased probability to release vesicles in response to a PSAP as a consequence of an increased Ca^{2+}-concentration in the presynaptic terminal. Consequently, the synaptic connection will be strengthened for a short time because of either an increase in size of the readily releasable pool of vesicles that contain neurotransmitter molecules or because of an increase in the amount of neurotransmitter molecules stored in the vesicles released in response to an AP. The type of synaptic enhancement present at a given synapse depends on its input: a single AP leads to facilitation, whereas a short train of consecutive APs generally causes augmentation while longer trains lead to potentiation [18], [19]. The inset in Figure 2.13 illustrates how paired pulse facilitation affects the PSCs in a way that two consecutive APs separated by a time Δt evoke PSCs with the second response larger than the first, while the amplitude of facilitation depends on the temporal structure of synaptic input. The opposed effect to STF is *short-term depression (STD)* which decreases the amplitude of PSCs and occurs due to a decrease in the probability to release vesicles in response to a PSAP as a consequence of a decreased Ca^{2+}-concentration in the presynaptic terminal [20].

Complementary to STP, synaptic plasticity lasting several minutes or longer is categorized as *long-term plasticity (LTP)*, an exclusively postsynaptic mechanisms which is found to require the binding of glutamate and glycine for the activation of the specific receptors responsible for LTP, so called NMDA receptors [21]. Similar to STP, LTP is differentiated into *long-term potentiation (LTP - not to be confused with long-term plasticity)*, and *long-term depression (LTD)*. Although oppositional in their effects on synaptic transmission, both LTP as well as LTD are induced by a rise in the intracellular Ca^{2+}-concentration $[Ca^{2+}]_i$ in the postsynaptic neuron. The brief activation of an excitatory pathway can produce LTD only at a minimum level of postsynaptic depolarization caused by a rise in postsynaptic $[Ca^{2+}]_i$. LTP on the other hand requires much stronger postsynaptic depolarization and consequently a much higher postsynaptic $[Ca^{2+}]_i$, hence, it is possible to first induce LTD and then LTP at the same synapse [22]. The amplitude of Ca^{2+}-surge is critical for the induction of LTD and LTP rather than the source of Ca^{2+} since the activation of voltage-gated Ca^{2+}-channels and ligand-gated Ca^{2+}-channels as well as the release from intracellular stores have been shown to contribute to the induction of both LTD and LTP. Figure 2.14 illustrates the dependence of whether LTD or LTP is induced on the postsynaptic membrane potential V_m. Note that there are two ranges of V_m within which synapses do not undergo any long-term modifications. A moderate rise in $[Ca^{2+}]_i$ leads to a predominant activation of phosphatases, while a stronger depolarization favorably leads to the activation of kinases, which has the opposing effect to the activation of phosphatases [24]. Phosphatases and kinases are both enzymes whose acti-

Chapter 2: A Biological Background

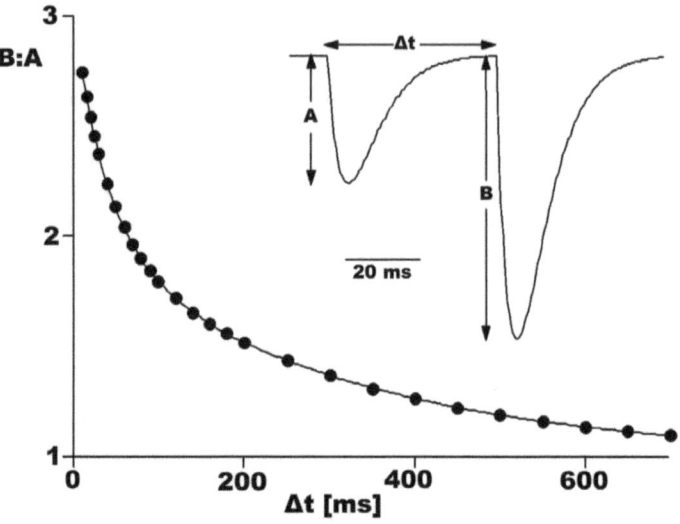

Figure 2.13.: Simulation of an experiment that shows paired pulse facilitation (PPF) which occurs at many synapses. As shown in the plot, the ratio of amplitudes of two consecutive PSCs being the results of two consecutive PSAPs depends on the temporal delay Δt between those two APs. The inset illustrates the growth in PSCs as a result of increased probability to release neurotransmitter. [modified from [18]]

vation is involved in the mechanisms of the transduction of signals in the nervous system but the exact understanding of the underlying mechanisms for the activation of these enzymes is not mandatory for the concepts of this thesis, thus, further explanations are left out but can be found elsewhere, e.g. in [13].

A specific form of LTP is *spike-timing-dependent plasticity (STDP)*, a mechanism which adjusts the connection strengths of synapses between neurons based on the relative timing of a particular neuron's output APs and input APs, also called spike timing. (In the context of neural computation, APs are called *spikes* to emphasize their all-or-nothing character and their treatment as single unitary events which encode information). The main feature of STDP is the fact that synapses increase their efficacy if the presynaptic AP arrives shortly before a postsynaptic AP is generated and in this way **STDP, with regards to LTP and LTD, is closely related to hebbian learning as an attempt to explain the development of an individual's brain on a cellular level.**

2.2. The Synapse

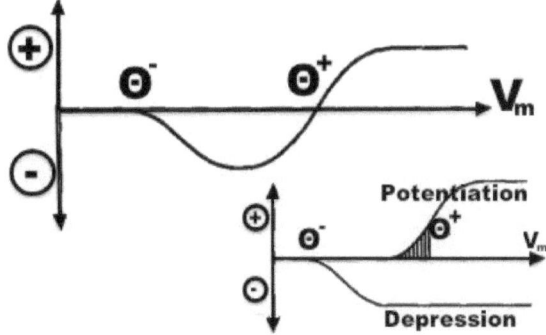

Figure 2.14.: The direction of synaptic weight change depends on the postsynaptic membrane potential V_m, i.e. on the amplitude of the surge of $[Ca^{2+}]_i$. If the first threshold Θ^- is reached, LTD is induced and if the second threshold Θ^+ is reached, LTP is induced (inset). [modified from [23]]

2.2.4. Spike-Timing-Dependent Plasticity (STDP)

STDP is a specific form of LTP which was explored in 1998 by Bi and Poo who investigated cultures of dissociated rat hippocampal neurons [25]. Their main result states that persistent

---
Info-box:
The hippocampus is a major component of the brain of vertebrates and belongs to the limbic system, a set of brain structures that support a variety of functions such as emotion, behaviour, motivation, long-term memory and olfaction. It plays an important role in the consolidation of information from short-term memory to long-term memory and spatial navigation. [26]

---

potentiation and depression of glutamatergic synapses can be induced by correlated spiking of presynaptic and postsynaptic neurons, whereas the relative timing between presynaptic and postsynaptic spiking determines whether the synaptic weight is increased or decreased. A key observation is that the synaptic weight of a synapse connecting a presynaptic neuron X with a postsynaptic neuron Y increases if the activities of both neurons gave reason to believe that there is a causal relationship between neuron X and neuron Y, i.e. if neuron X temporally fires before neuron Y does, therefore being the cause of activity of neuron Y. On the contrary, if neuron Y temporally fires before neuron X does, the synaptic weight would

Chapter 2: A Biological Background

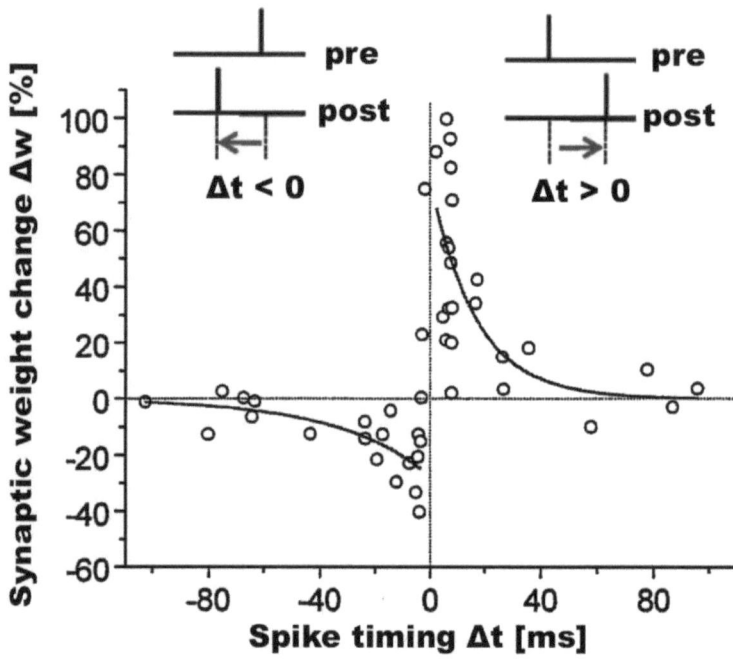

Figure 2.15.: The synaptic weight changes depending on the relative spike timing between presynaptic and postsynaptic neuron. The synaptic weight change represents the change in EPSC amplitude which was measured 20-30 minutes after the repetitive correlated spiking (60 pulses at 1 Hz). The inset illustrates the temporal structure of presynaptic and postsynaptic neuronal activity for both cases, $\Delta t > 0$ and $\Delta t < 0$. [modified from [25]]

decrease since it is unlikely that the activities of both neurons have any causal relationship. Figure 2.15 illustrates the synaptic weight change Δw of a synapse connecting two neurons which depends on the relative spike timing Δt of both neurons. For positive values of Δt, LTP is induced which only occurred at synapses with low initial connection strengths, whereas for negative values of Δt, LTD is induced which didn't show any obvious dependence on the initial synaptic strength. The effect of STDP was quantified amongst others by Wong et al. by fitting the biological data using a simple exponential:

$$\Delta w = A \cdot e^{\Delta t/\tau}$$

where A and τ are two free parameters that correspond to the scaling factor and the time constant for STDP curves, respectively. The exponential fit is very convenient to formalize STDP into a simple parametric model that is often used in computational studies. Recent experiments have shown that the time constant of the STDP curve shows significant variations depending on the synapse's location in the brain: e.g. hippocampal glutamatergic synapses show potentiation with $\tau = 16.8$ ms while synapses connecting neurons in the visual cortex show potentiation with $\tau = 13.3$ ms. It is commonly believed that synapses with different values for τ may serve specific functions in information processing at different stages of neural pathways [5].

2.3. An Overall View

The superiority of the human brain to the brain of other animals results from the ultrahigh density of main signaling units (neurons) communicating with each other via synapses which allows to receive and process information in a vastly parallel fashion. Emergent phenomena such as behaviour, memory and learning result from the interplay of a huge number of these simple components. The neural pathways through which informational content, encoded in APs, is transported can be modified depending on the informational input provided by the environment in which the organism controlled by the brain operates. More precisely, synapses undergo functional and structural changes depending on neuronal activity in order to optimize the flow of information in the nervous system, thus, generating a desired response to the stimuli to which the organism is exposed. **The mechanisms of synaptic plasticity are commonly believed to be the underlying mechanisms for emergent phenomena mentioned above and thus, recent research is based on the idea to emulate the biological functionalities of synaptic plasticity.**

Chapter 2: A Biological Background

CHAPTER 3

Experimental Emulations

The two main goals of experimental emulations of the functionalities of biological neural networks (BNNs) are: (1) to gain a better understanding of how the brain works, especially in terms of remarkable phenomena such as memory and learning. This understanding would result from new accessible possibilities to extensively research these emulated functionalities in artificial neural networks (ANNs); (2) to utilize these functionalities to overcome the limitations of today's computational standards which are based on binary logic. One of the first steps of experimental emulations is based on common CMOS technology which has the big advantage of being well understood due its successful implementation in many different fields for many decades. However, its downside is that the emulation of biological functionalities is constructed in a completely artificial fashion and in no way based on any inherent physical behaviour of a specific component which would serve as an analogy to a biological role model. More recent attempts to emulate biological functionalities, especially synaptic plasticity, utilize the inherent physical characteristics of specific materials that provide the possibility to influence the electrical resistance and thus, the conductivity of a device which contains said specific material, in a more or less continuous fashion.

There are several candidates being under research for synaptic plasticity emulation purposes: (1) the memristor, the fourth basic two-terminal device amongst the common resistor, the capacitor and the inductor, shows a static relationship between the charge q and the flux ϕ [10]. Flux driven memristors behave as resistors whose resistance depends on the flux ϕ, i.e. the integral over time of the applied voltage to the device. Their resistance can be gradually changed by controlling the time over which a certain voltage is applied to the device, thus, making it a suitable candidate to serve as an artificial synapse [9–12]; (2) material compositions that are able to exhibit the growth and destruction of some sort of conducting filament upon voltage appliance between two electrodes confining this composition. This

Chapter 3: Experimental Emulations

conducting filament provides the possibility to switch from a low conductive state into a high conductive state upon connecting both electrodes with each other and vice versa upon subsequent disconnection of both electrodes. Recently researched compositions are e.g. AgS_2 which forms conducting silver bridges that spontaneously decay under certain conditions providing the possibility to emulate STP and LTP, resembled through the formation of a spontaneously decaying silver bridge and a long lasting silver bridge, respectively [8]. Another class of suitable compositions are metal oxides such as HfO_x and AlO_x which form conducting bridges of oxygen vacancies upon positive voltage appliance that can be destroyed again upon negative voltage appliance providing the possibility to emulate STDP as a specific form of LTP in a device containing such compositions serving as an artificial synapse [6]. (3) Phase-change materials that are able to undergo a structural change accompanied by a dramatic change in conductance which makes devices containing such materials a suitable solution for an artificial synapse [5]. **This chapter presents the reader with examples of experimental emulations based on CMOS technology as well as on phase-change materials.**

3.1. Modeling STP and LTP in a CMOS Spiking Neural Network Chip

The implementation of synaptic plasticity based on CMOS spiking neurons is motivated by the increased consensus within the neuroscience community considering single neuron models and the modeling of synaptic transmission including the different aspects of plasticity [27]. The study of the different aspects of synaptic plasticity involves timescales ranging from milliseconds to hours for STP and LTP, respectively. This temporal dynamic range can be made accessible by highly accelerated analog VLSI (Very Large Scale Integration) models of leaky integrate and fire neurons (see also section 4.4), which incorporate fast and slow synaptic facilitation and depression mechanisms in conductance based synapses. A single chip of $25\,mm^2$ containing 10^5 synapses and 384 neurons can model the temporal evolution of the synaptic weights with an acceleration factor of 10^5, i.e. ten thousand times faster than biological real time evolution, which reduces the time needed to model ten minutes of biological activity to merely 6 ms, allowing complex parameter searches to reproduce biological findings.

Two mechanisms of synaptic plasticity modulating the same parameter on completely different timescales are implemented: (1) STP based on the history of presynaptic APs emulating the limitation of resources involved in the synaptic transmission, e.g. neurotransmitters and (2) STDP based on the precise timing between pre- and postsynaptic APs emulating the long-term temporal evolution of the artificial neural network. A Neuron's membrane potential

3.1. Modeling STP and LTP in a CMOS Spiking Neural Network Chip

$V(t)$ is described by the leaky I&F model (see also section 4.4):

$$C_m \frac{dV}{dt} = g_m(V - E_l) + \sum_j p_j(t) g_j(t)(V - E_x) + \sum_k p_k(t) g_k(t)(V - E_i),$$

where C_m represents the total membrane capacitance. The first term on the right side models the contribution of the resting channels, i.e. E_l is the membrane potential value at rest. The sums over j and k run over all excitatory and all inhibitory synapses, respectively, using different reversal potentials E_x and E_i, whereas the individual activations of synapses are controlled by the synaptic open probability $p_{j,k}(t)$. Synaptic plasticity is included by the variation of excitatory and inhibitory conductance g_j and g_k with time. The synaptic conductance $g_{j,k}$ is modeled as the product of the synaptic weight $\omega_{j,k}(t)$ and a maximum conductance $g_{j,k}^{max}(t)$:

$$g_{j,k}(t) = \omega_{j,k}(t) \cdot g_{j,k}^{max}(t).$$

Synaptic conductance is modified by both, STP and LTP, whereas STP acts by temporarily increasing or decreasing the maximum conductance $g_{j,k}^{max}(t)$, while LTP, i.e. STDP, modifies the synaptic weights $\omega_{j,k}(t)$. STP is implemented by the distribution of the absolute synaptic efficacy A_{SE} between an inactive partition (I) and a recovered partition (R):

$$R = 1 - I.$$

In case of STD, each AP generates a conductance pulse with g^{max} being proportional to the percentage of total efficacy momentarily assigned to the recovered partition (R):

$$g^{max} = A_{SE} \cdot R.$$

After the AP is transmitted to the postsynaptic neuron, a fixed fraction of the recovered efficacy, the usable synaptic efficacy U_{SE}, is transferred to the inactive partition (I). This transfer is repeated for each AP, while the inactive partition (I) loses efficacy to the recovered partition (R) by a time-continuous recovery process:

$$\frac{dI}{dt} = -\frac{I}{t_{rec}} + U_{SE} \cdot R \cdot \delta(t - t_{AP}).$$

For the implementation of STF instead of STD, the roles of I and R are exchanged and g^{max} becomes proportional to I. The correlation measurement for STDP as a specific form of LTP is implemented at every single synapse, where the time Δt between a pre- and postsynaptic AP is measured. The synaptic weight change depends on the the measured value of Δt is is modeled by:

$$\omega_{new} = \omega_{old} \cdot (1 + F(\Delta t)),$$

Chapter 3: Experimental Emulations

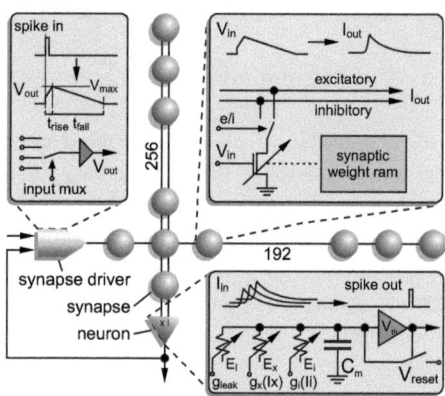

Figure 3.1.: *Operation principle of the artificial neural network chip. A synapse array consists of 256 rows and 192 columns of synapses with 192 neurons at the bottom end of each column receiving input from 256 synapses, one in each row. The neuron integrates all EPSCs and IPSCs and decides whether a spike is generated or not depending on whether V_m crosses V_{th} or not. V_m is reset to V_{reset} upon the generation of a spike. The spikes of all 192 neurons are transmitted to the 256 synapse drivers at the beginning of each row, where the input spikes are converted into a voltage ramp. This voltage ramp serves as synaptic input where it is converted into an output current, which can be either excitatory or inhibitory. The synaptic weight is stored in a static RAM in each synapse. The synaptic output currents are transferred to the neuron where the cycle of the artificial neural network is closed. [taken from [27]]*

where $F(\Delta t)$ is the STDP modification function which is defined according to its biological role model:

$$F(\Delta t) = \begin{cases} A_+ exp\left(-\frac{\Delta t}{\tau_+}\right), & \text{if } \Delta t > 0 \\ -A_- exp\left(\frac{\Delta t}{\tau_-}\right), & \text{if } \Delta t < 0 \end{cases},$$

where A_+, A_-, τ_+ and τ_- are fit parameters that can be experimentally determined.

The network model is based on the biological principles of the information processing architecture of the brian. Each neuron can be connected to any other neuron, whereas the number of synapses per neuron is limited by the chip to 256, thus, each output neuron has a subset of input neurons. Figure 3.1 illustrates the operation principle of the network chip. The chip contains two synapse arrays with about 50k synapses per array which occupy most

3.1. Modeling STP and LTP in a CMOS Spiking Neural Network Chip

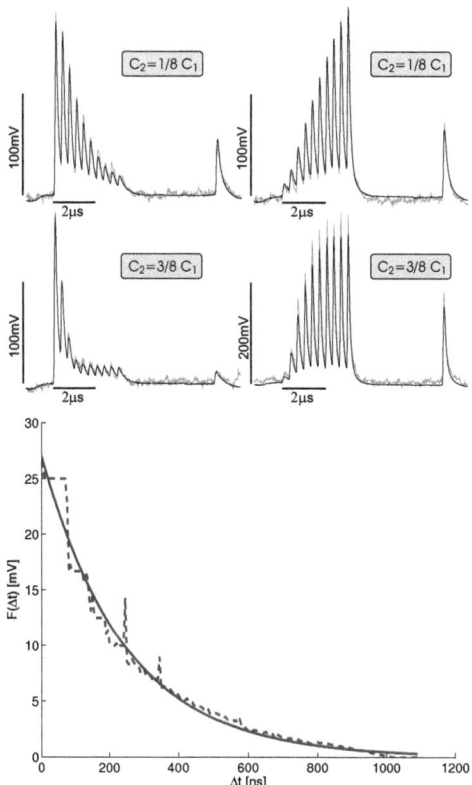

Figure 3.2.: *Illustration of the measurements for STF and STD as well as STDP.* **Top:** *Measurement of the postsynaptic membrane potential V_m showing STD on the left and STF on the right. The top and bottom traces differ by the ratio of C_1 to C_2 which determines to what extent STD and STF is modeled at each synapse. The grey traces show single measurements while the solid traces show the averaged results from 500 runs.* **Bottom:** *Measurement of the STDP modification function $F(\Delta t)$ for a single arbitrary synapse (dashed line). The measurement is in good agreement with the theoretical model shown as a reference (solid line). [taken from [27, 28]]*

of the chip's area. Both arrays consist of 256 rows and 192 columns with one neuron at the bottom end of each column. Each neuron contains an artificial membrane capacitance C_m

Chapter 3: Experimental Emulations

and three different conductances, modeling the resting ion channel currents, the EPSCs and the IPSCs, as well as the threshold and reset voltages V_{th} and V_{reset}. Whether synapses are excitatory or inhibitory is determined by the synapse driver at the beginning of each row of synapses which can switch the synapses output between its excitatory or inhibitory input line. The synaptic weight is stored in static RAM whose content is converted into a current. The top part of Figure 3.2 illustrates the measurement of the postsynaptic membrane potential V_m showing STD and STF. The ratio of two capacitances C_1 and C_2 to each other allows to modulate to what extent STD and STF is modeled at each synapse. Note the much shorter time scale on which STP is implemented in the artificial synapses in comparison to biological STP (see Figure 2.13). The bottom part of Figure 3.2 illustrates the measurement of the STD modification function for a single arbitrary synapse. During the measurement, pairs of pre- and postspikes with a time difference Δt between them were sent into the chip, whereas ten measurements were recorded for each time difference Δt which was incremented by 5 ns after each data point.

In summary, an artificial neural network chip has been presented which implements STD and STF as forms of STP as well as STDP as a specific form of LTP. Although the measurements are in good agreement with theoretical models and show great similarity to biological recordings, the space of this emulation of a neural network is quite large considering that each synapse contains 76 MOS transistors. **The next section presents an architecture based on an electronic device serving as an artificial synapse which implements STDP based on inherent physical processes within the device itself, rather than on electrical engineered circuitry. This gain in space is a crucial step towards reaching the ultrahigh density of synapses of the brain.**

3.2. Implementation of STDP based on Phase-Change Material Synapses

Since synaptic plasticity, especially STDP with respect to hebbian learning, is commonly believed to be closely related to phenomena such as memory and learning, interest in emulating synaptic plasticity based on different devices serving as a nano scalable artificial synapses has rapidly grown. One promising candidate for such a device are phase-change materials providing advantageous properties in terms of scalability, energy consumption and endurance. **Preparatory to the presentation of a phase-change material crosspoint structure emulating STDP, a short introduction of phase-change materials and their electrical switching behaviour is presented.**

3.2. Implementation of STDP based on Phase-Change Material Synapses

3.2.1. Phase-Change Materials

The common motivation for research of phase-change materials (PCM) is the desire to create a universal memory. While most research was originally driven by the need of phase-change memory to fulfill the requirements of a universal memory, namely non-volatility, high storage density, fast read, write and erase speed and low power consumption [29], recent research is based on the idea to utilize the inherent physics of phase-change materials to emulate biological functionality and especially synaptic plasticity [5, 30].

One inherent characteristic is the phase transition behaviour of PCM cells that can be induced by optical and electrical excitation. This phase transition has been demonstrated in a proof-of-principle experiment to be useful for the realization of a processing unit which simultaneously performs processing and memory functions at the same location, a feature which closely resembles the operation principle of brain-like systems and thus, happens to be a promising candidate to achieve dramatic improvements in performance [31]. In this experiment, the execution of the four basic arithmetic processes with simultaneous storage of the result has been demonstrated by utilizing the PCM's natural accumulation property which can be also exploited for the implementation of a PCM-I&F neuron [32, 33] (see section 4.4). The accumulation property of $Ge_2Sb_2Te_5$ was measured in a femtosecond laser experiment in which the optical reflectivity of the PCM demonstrated a non-linear increase - caused by the phase transition from the amorphous state into the crystalline state - depending on the number of applied laser pulses, thus facilitating the possibility of arithmetic calculation and simultaneous storage of the results at the same location.

Another possibility to induce the phase transition is the electrical excitation of the PCM. Under the influence of an electrical current, PCM's undergo a structural transformation between the amorphous and crystalline phase, whereas the amorphous phase, contrary to the crystalline phase, tends to have a high electrical resistivity, sometimes three or four orders of magnitude higher [34]. The transition from the amorphous state into the crystalline state, called set operation, is induced by applying a voltage pulse to the phase-change cell to heat a significant portion of the material above the crystallization temperature. The transition from the crystalline state into the amorphous state, called reset operation, is induced by applying a much larger voltage pulse to melt the central portion of material inside the cell. This pulse must be cut off abruptly in order to quench the molten material into the amorphous phase. The read operation simply measures the cell's resistance at small voltages so that the device state is not perturbed. Figure 3.3 shows a cross-section schematic of a conventional PCM cell and summarizes the three basic programming operations. Due to the phase transition during electrical switching, the current-voltage (I-V) characteristics of a PCM cell is not purely ohmic and depends on the state of the cell. Figure 3.4 shows I-V curves of the set and reset state. While the I-V characteristics for the crystalline state shows ohmic behaviour, this is not the

Chapter 3: Experimental Emulations

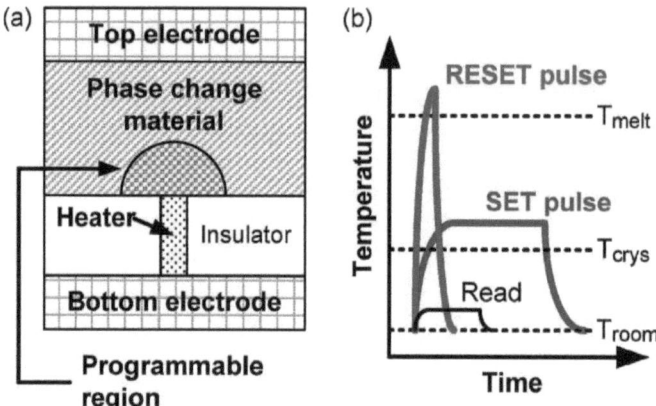

Figure 3.3.: *Programming of a PCM cell is done with voltage pulses. **a)** Cross-section schematic of a conventional PCM cell. Electrical current passes through the phase-change material between the top electrode and the heater. The programmable region results from current crowding at the heater-PCM interface. **b)** Three basic voltage pulses are needed for programming which change the temperature of the phase-change material accordingly. [taken from [34]]*

case for the amorphous state. In the amorphous off state (a-off state), almost no current flows through the device for applied voltages V_A below the threshold switching voltage V_{th}. At V_{th}, the device switches from the a-off state into the amorphous on state (a-on state) after a delay time τ_d showing a negative differential resistance (NDR). This switching behaviour is reversible if the applied voltage pulse is removed very quickly. For voltage pulses applied long enough, the cell remains in the a-on state and is eventually heated above the crystallization temperature which leads to the phase transition [34, 35].

In addition to the features discussed above, transient effects can be observed by examining the time dependent behaviour of voltage and current [36], however, the understanding of these effects is not mandatory for the following section but will be explained later in chapter 5, section 5.3.

The underlying mechanisms of the threshold switching - the switching effect from the a-off into the a-on state - in PCM cells, have been described by three models which are tried to be verified or falsified at the present time of research [82, 86, 87]. Ielmini et al. proposed a model which explains the threshold switching mechanism by the non-equilibrium pop-

3.2. Implementation of STDP based on Phase-Change Material Synapses

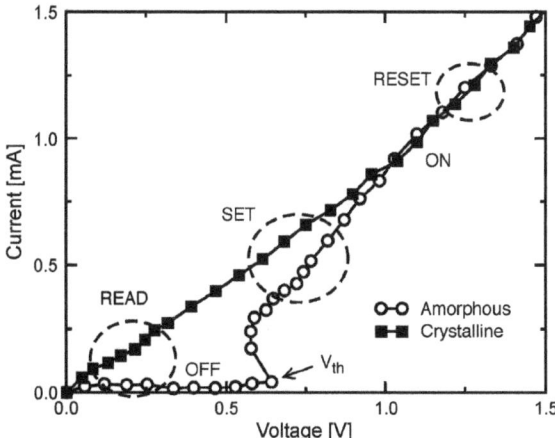

Figure 3.4.: *I-V characteristics of set and reset state. The I-V characteristics of the crystalline state shows ohmic behaviour, while the amorphous state shows switching behaviour at the threshold voltage V_{th} where the material reversibly switches from the high resistive a-off state into the low resistive a-on state. [taken from [34]]*

ulation in high-mobility shallow traps at high electric field and by the non-uniform field distribution along the amorphous layer thickness. A single analytical model was derived which accounts for effects such as subthreshold conduction, threshold switching, NDR region and the a-on state regime [87]. Pirovano et al. proposed an original band gap model which is consistent with the microscopic structure of both, crystalline and amorphous chalcogenide accompanied by a physical explanation of the electronic threshold switching in PCMs linking the characteristic effects to the competing role of impact ionization and recombination via valence alternation paris [86]. Karpov et al. proposed a simple model of threshold switching in PCMs based on the field induced nucleation of conductive cylindrical crystallites which is able to explain the correlations between the threshold switching voltage and material parameters, e.g. the nucleation barrier and radius, amorphous thickness as well as the dependence of the threshold voltage on the temperature and the delay time [87]. **Even though until today, none of the above models turned out to be the only true model (however, none of them has been disproven either) and thus, the threshold switching is not yet completely understood, the electrical switching behaviour has been utilized for experimental emulations of biological functionalities as presented in the next section.**

Chapter 3: Experimental Emulations

3.2.2. A Phase-Change Cross-Point Structure emulating STDP

In 2012, Wong et al. presented a concept which utilizes the continuous transition between resistance levels of PCM in an analog manner to emulate biological synapses, as illustrated in Figure 3.5. The PCM synapses consist of $Ge_2Sb_2Te_5$ (GST) deposited between a tungsten

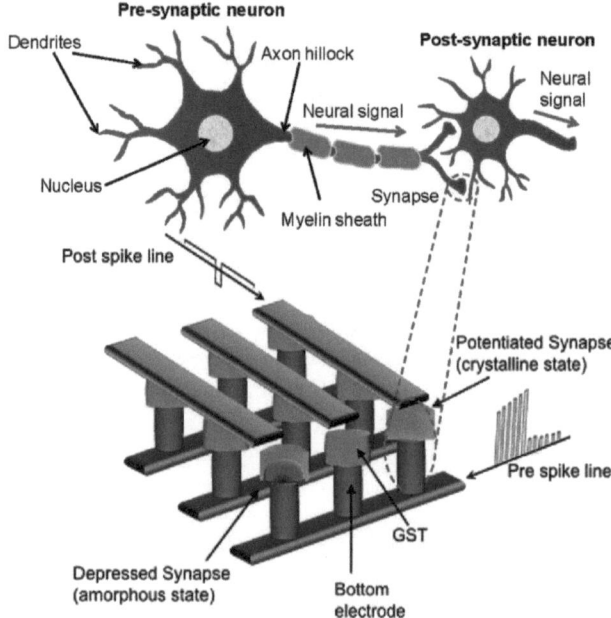

Figure 3.5.: Interconnection scheme of PCM synapses in a 3D stackable cross-point structure to reach ultrahigh density and compactness of the brain. In the cross-point structure, prespike and postspike electrodes are connected by several PCM synapses in analogy to its biological role model. The synaptic weight of the artificial PCM synapses is represented by the resistance level of the PCM device, whereas the amorphous state represents a depressed synapse while the crystalline state represents a potentiated synapse. The synaptic weight can be changed by applying specific voltage pulse at the PCM device at the pre- and postspike line. [modified from [5]]

bottom electrode with a small contact area diameter of 75 nm capped by TiN and a TiN top electrode. The most important requirement of an electronic device serving as an artificial

3.2. Implementation of STDP based on Phase-Change Material Synapses

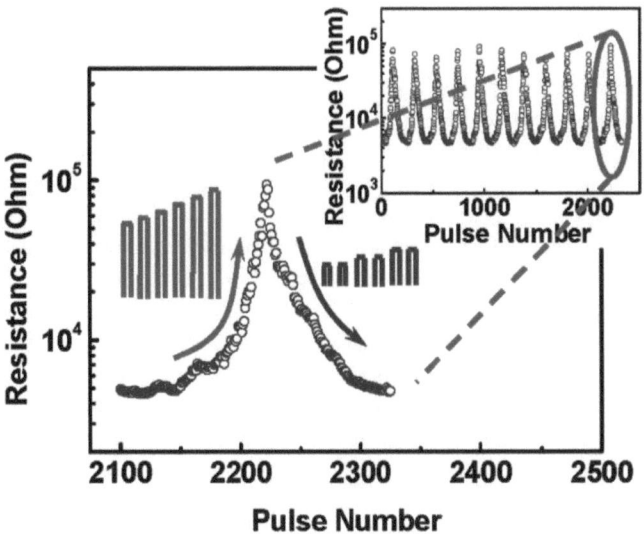

Figure 3.6.: *The continuous change in synaptic weight can be emulated by a fine control of the PCM cell's resistance. Gradual reset of the cell is implemented by using voltage pulses with increasing amplitudes between 2-4 V and 20 mV increase per step. The pulse width, rise and fall times are 75 ns, 25 ns and 25 ns, respectively. Gradual set of the cell is implemented by using stair case voltage pulses with increasing amplitudes between 0.5-0.9 V and 0.1 V increase per 20 steps. The pulse width, rise and fall times are 5 μs, 500 ns and 500 ns, respectively. The inset illustrates the reproducibility of the gradual change in resistance. [modified from [5]]*

synapse is the emulation of the analogue nature of synaptic weight change in biological synapses [25] which is fulfilled by the continuous transition between intermediate resistance states of the PCM, as illustrated in Figure 3.6. A very fine control of resistance, close to 1% change per synaptic event - i.e. an order of magnitude change in the PCM cell's resistance is achieved through 100 steps for both, set and reset transitions - can be achieved by gradually increasing voltage pulses with a custom engineered timing to probe the intermediate resistance levels and to maintain almost continuous transitions between adjacent resistance levels. Gradual reset of the cell is implemented by using voltage pulses with increasing amplitudes between 2-4 V and 20 mV increase per step. The pulse width, rise and fall times are 75 ns, 25 ns and 25 ns, respectively. Gradual set of the cell is implemented by using stair case voltage

Chapter 3: Experimental Emulations

pulses with increasing amplitudes between 0.5-0.9 V and 0.1 V increase after 20 steps, i.e. 20 pulses of each voltage value are applied before the increase to the next voltage value. The pulse width, rise and fall times are 5 µs, 500 ns and 500 ns, respectively.

In order to utilize this fine control of resistance for the emulation of STDP, which depends on the precise timing of pre- and postspike, a specifically engineered pre- and postspike signal is applied to the top electrode and bottom electrode of the PCM synapse, respectively. The spike signals have no similarity to biological APs and are developed based on the pulsing scheme used in gradual set and reset experiments with GST cells. Figure 3.7 illustrates the STDP learning curve of biological data [25] and the STDP learning curve recorded for the PCM synapse, as well as a simplified schematic of the spike signals which is used to emulate STDP. The prespike signal consists of a pulse train of depression pulses, corresponding to the reset state of the PCM cell, with increasing amplitudes followed by a pulse train of potentiating pulses, corresponding to the set state of the PCM cell, with decreasing amplitudes. The total duration of the prespike signal is 120 ms, whereas the depression pulses width, rise and fall times are 50 ns, 10 ns and 10 ns, respectively, while they are 1 µs, 100 ns and 100 ns for potentiating pulses with a constant time spacing of 10 ms between two consecutive pulses. The postspike signal, serving as a gating function, is a low amplitude, continuous pulse with 120 ms duration and contains a short negative amplitude pulse of 8 ms at the center. The difference between both signals $V_{pre} - V_{post}$ defines the net programming voltage across the PCM cell at each point of time. Hence, the time difference Δt between pre- and postspike signal determines which of the prespike pulses overlaps with the postspike pulse, thus, determining whether a depression pulse or potentiating pulse corresponding to an increase or decrease of the cells resistance, respectively, is applied to the PCM cell. For positive values of Δt, a potentiating pulse is applied which decreases the cell's resistance resembling an increase of synaptic weight, whereas for negative values of Δt, a depressing pulse is applied which increases the cell's resistance resembling a decrease of synaptic weight. The overall STDP curve measured by repeating the spike scheme for different values of Δt is in good agreement with the biological data measured in hippocampal glutamatergic synapses by [25]. Further experiments showed, that different STDP learning curves can be obtained through the modification of the amplitudes and the time spacing between individual pulses in the prespike signal [5].

In summary, this work presented how phase-change materials can be used to serve as artificial synapses in a cross-point structure that emulate STDP which is found to be exhibited by biological synapses in the many different regions of the brain [39]. A specifically engineered pulse sequence is used to implement the change in the cell's resistance which resembles the change in synaptic weight depending on the precise timing between pre- and postspike signals. The downside of this work with respect to the desire to mimic biology is that the utilized pre- and postspike signals have no resemblance to biological APs whatsoever.

3.2. Implementation of STDP based on Phase-Change Material Synapses

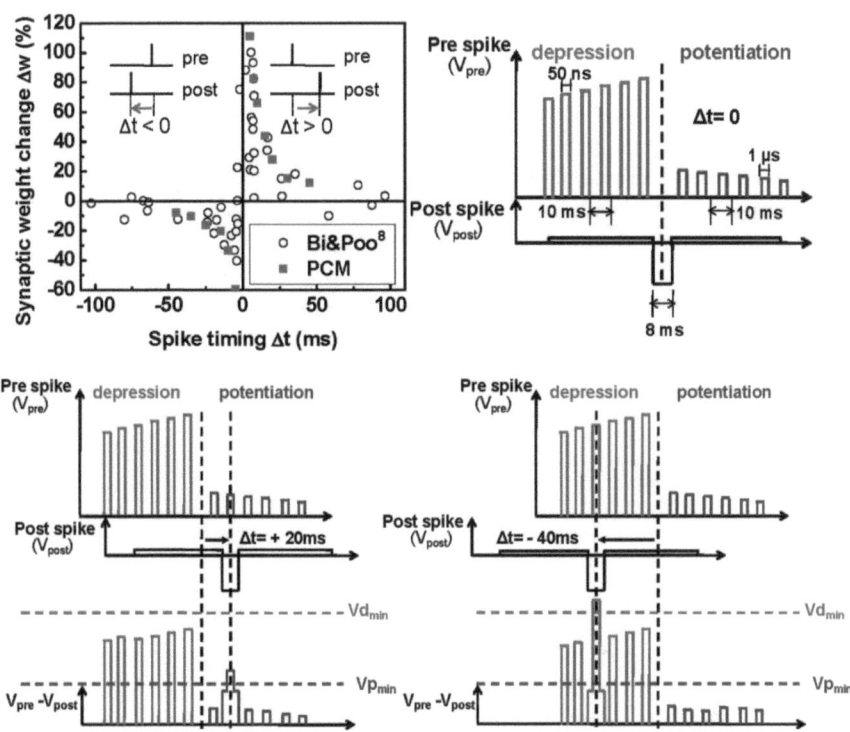

Figure 3.7.: Implementation of STDP through a specifically engineered pulse sequence. **Top left:** The STDP learning curve obtained for the PCM synapses (red squares) is in good agreement with biological data recorded in hippocampal glutamatergic synapses by [25]. **Top right:** Specifically engineered pre- and postspike signal which is applied to the top and bottom electrode of the PCM cell, respectively. **Bottom left:** For positive values of Δt, the overlap of pre- and postspike signal results in a potentiation pulse which decreases the PCM cell's resistance. **Bottom right:** For negative values of Δt, the overlap of pre- and postspike signal results in a depression pulse which increases the PCM cell's resistance. [modified from [5]]

The general idea to utilize the large resistance contrast of PCM for artificial neural networks, or more precisely for neuronal behaviour, has been patented in 2006 [40] an will be briefly dealt with in the next section.

Chapter 3: Experimental Emulations

3.3. Phase-Change Materials for Artificial Neural Networks

The cumulative nature of artificial neurons based on silicon emulating the functionalities of biological neurons is based on conventional binary processing. In 2006, Stanford Ovshinsky, an american inventor and scientist, had the idea patented to utilize the large resistance contrast of PCM and the inherent characteristics to respond to a plurality of signals for artificial neural networks. Figure 3.8 illustrates a schematic of an artificial neuron with three basic elements: (1) the weighting unit which receives an input signal and transmits this signal with a specific relative weight ω; (2) the accumulation unit which accumulates the energy of the weighted signals up to a threshold amount of energy, whereupon the accumulation unit is capable of generating an output signal; (3) the activation unit which modifies the output signal generated by the accumulation unit according to a mathematical objective. All three

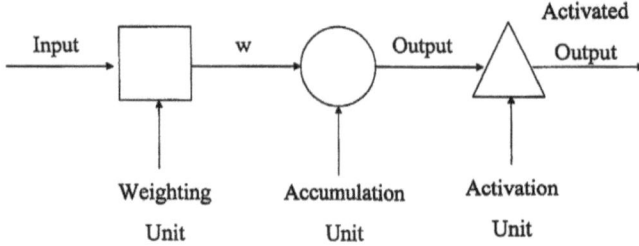

Figure 3.8.: *Schematic Illustration of an artificial neuron with three basic elements: (1) the weighting unit which receives an input signal and transmits this signal with a specific relative weight ω; (2) the accumulation unit which accumulates the energy of the weighted signals up to a threshold amount of energy, whereupon the accumulation unit is capable of generating an output signal; (3) the activation unit which modifies the output signal generated by the accumulation unit according to a mathematical objective. [taken from [40]]*

elements are suggested to contain an instant neural computing medium which comprises a PCM that is able to cumulatively respond to multiple input signals and a large resistance contrast. The weighting function of the weighting unit is suggested to be emulated by the PCM's large resistance contrast, e.g. as presented in the previous section. The change in weight corresponds to a change in resistance, whereas the ability of the weighting unit to transmit signals is modulated by its resistance. A high weight corresponds to a low resistance, whereas a low weight corresponds to a high resistance. The accumulation unit's functionality is suggested to be emulated by the cumulative nature of the instant neural computing medium, more precisely by the ability of the PCM to switch from a high resistance reset state into a low resistance set state upon the accumulation of energy. The threshold firing, i.e. the generation of an output signal, corresponds to said switching event, which is accompanied by a dramatic increase in conductivity facilitating the transmission of an output signal to other artificial neurons.

In summary, Ovshinsky proposed a general idea of how to use PCM for neural networks based on its inherent physical characteristics. The artificial neuron utilizes the large resistance contrast as well as the electrical switching behaviour between set and reset state of a PCM comprised by an instant neural computing medium. [40]

3.4. An Overall View

Emulations of the functionalities provided by a biological computing machine, the brain, are based on several different technologies. A CMOS based chip has been presented which contains a whole artificial neural network emulating STP as well as STDP as a specific form of LTP. The goal to reach the ultrahigh density in synapses of the brain is limited by the space taken away by a single CMOS synapse, thus new concepts for artificial synapses that are much smaller are being researched in order to reach that goal. In general, electronic devices that are suitable to serve as an artificial synapse are required to possess a characteristic which can be more ore less gradually changed and stored in a non volatile way in order to emulate synaptic weight. Amongst others, promising candidates for this task are memristors, metal oxides, magnetic tunnel junctions and phase-change materials. In this chapter, a concept for a 3D stackable cross-point structure with PCM synapses has been presented which successfully implemented STDP. Besides the utilization for synaptic purposes, PCMs were also proposed by Ovshinksy to be useful for neuronal purposes. **Preparatory to the discussion of a more concrete concept of a PCM bursting neuron which picks up on Ovshinsky's general idea, the next chapter introduces bursting neurons, a specific class of neurons in the human brain.**

Chapter 3: Experimental Emulations

CHAPTER 4

Bursting Neurons

Some neurons in the human brain are capable of exhibiting a specific firing pattern, called *bursting*, which is thought to play an important role in communication between neurons [41–45]. Bursting is a dynamic state in which a neuron repeatedly fires discrete groups of consecutive AP's, also referred to as *bursts*. Two consecutive bursts are separated by a period of quiescence after which the neuron can send out another burst. **In order to emulate this specific behaviour, its underlying physiological mechanisms have to be examined.**

4.1. Physiological Mechanisms of Bursting

Neuronal bursting activity can occur as a result of inhibitory feedback used by sensory systems as a mechanism to 'toggle' between oscillatory (bursting) and non-oscillatory firing states, depending on the spatiotemporal structure of the input [46]. Weakly electric fish offer a clear example because prey and communication signals are significantly different from each other in their spatial extent. The electric fish *Apteronotus leptorhynchus* produces an electric field through rhythmic electric organ discharges (EOC) whose amplitude modulations caused by conspecifics or objects can be detected by electroreceptor afferents. These afferents encode amplitude modulations produced by both, prey and communication signals and synapse them onto electrosensory lateral line lobe (ELL - the nucleus that receives direct input from peripheral electroreceptor afferents) pyramidal neurons [47, 48], whereas prey signals are spatially localized [49] and communication signals, produced by conspecifics, are spatially diffuse [50]. *In vivo* recordings showed that these pyramidal neurons distinguish between prey-like and communication-like stimuli by producing oscillatory or non-oscillatory responses, respectively. Figure 4.1 illustrates the spatiotemporal structure of both forms of input and their dispersion on the skin of the fish as well as the corresponding *interspike interval (ISI)*

Chapter 4: Bursting Neurons

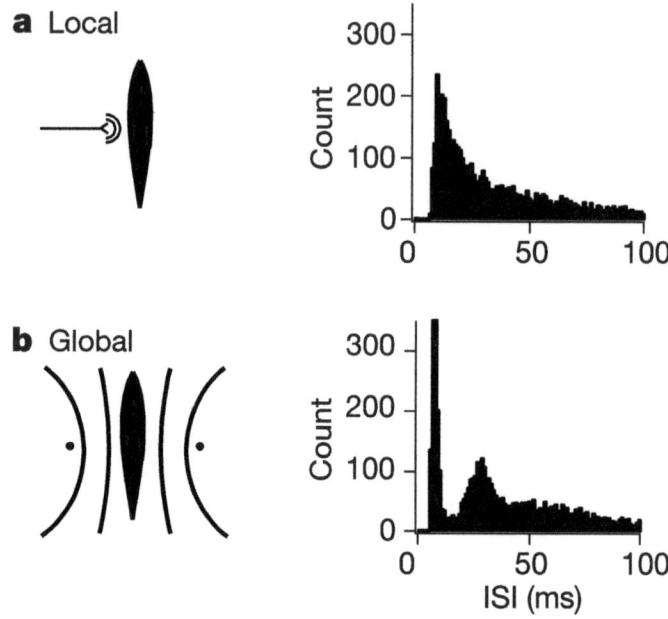

Figure 4.1.: ELL pyramidal neurons show differential responses to local (prey-like) and global (communication-like) stimuli. **a)** Local stimulation induced by a dipole placed near the skin of the fish to stimulate only a part of the pyramidal cell receptive field centre. The corresponding ISI histogram shows only one peak which stands for the interspike frequency of the pyramidal neurons. **b)** Global stimulation induced by two wire electrodes placed transverse to the fish producing spatially extensive stimuli. The corresponding ISI histograms shows two peaks which stand for the interburst and intraburst frequency. [modified from [46]]

histograms. Local and global input leads to non-bursting and bursting activity, respectively, which is encoded in the peaks of the ISI histograms. The two peaks in case of global input show intraburst and interburst intervals whereas the peak of the interburst interval is missing in case of local input. Based on well characterized circuitry [47], Doiron et al. modeled pyramidal neurons of the ELL as a layer of leaky integrate-and-fire neurons (see [46, 51] and also section 4.4), each connected to a central cell population $G(t)$ which feeds back an inhibitory synaptic response to all pyramidal neurons in the Pyramidal Cell Layer (PCL) after a fixed time delay

4.1. Physiological Mechanisms of Bursting

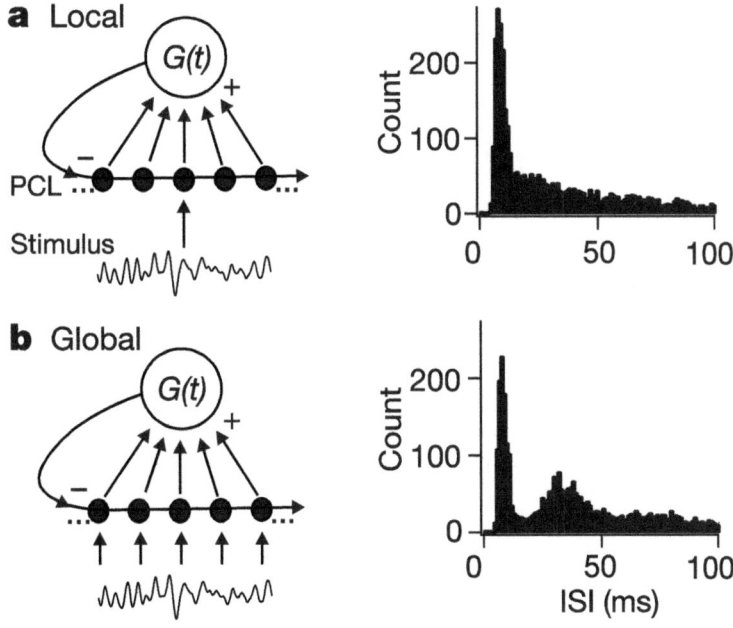

Figure 4.2.: *ELL neural network simulations involving global inhibitory feedback show differential responses to local and global stimuli. **a)** In case of local stimulation, only one pyramidal neuron of a pyramidal cell layer (PCL) receives the stimulus (schematically illustrated with a single arrow below the PCL) and projects excitatory spikes (illustrated through the + sign) to a cell population G(t) which feeds back inhibitory responses (illustrated through the - sign) to the PCL after a delay τ_d. **b)** In case of global stimulation, all pyramidal neurons of the PCL receive the stimulus (schematically illustrated with several arrows below the PCL) and project their spikes to G(t) which feeds back inhibitory responses to the PCL. Both corresponding ISI histograms are in good agreement with the in vivo recordings seen in Figure 4.1. [modified from [46]]*

τ_d for each spike it receives, as illustrated in Figure 4.2. The physiological interpretation of $G(t)$ is the integration of pyramidal cell output by bipolar cells and their GABA-mediated projection back to the ELL. The corresponding ISI histograms are in good agreement with

Chapter 4: Bursting Neurons

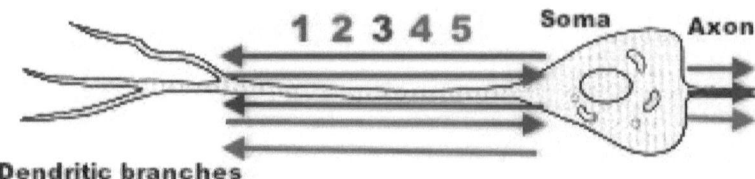

Figure 4.3.: *Bursting activity can result from the somatic-dendritic interplay. The colored arrows show the direction in which a spike is transported while the colors show the temporal order in which the spikes are generated. In some neurons, a somatic spike, i.e. an AP generated at the axon hillock, spreads not only along the axon but also across the whole soma and eventually reaches the dendrites (red arrow) where it evokes a dendritic spike in response which travels back to the soma (purple arrow). If the dendritic response of all excited dendrites is strong enough to generate another AP the process starts again and another spike spreads along the axon as well as across the soma and eventually reaches the dendrites (blue arrow) where another dendritic spike is evoked which travels back to the soma (purple arrow). Again, if the dendritic response is strong enough another AP is generated (green arrow) and so fourth. In this way, the neuron enters a bursting state. Note that the dendritic response represented by the purple arrows is the sum of all dendritic potentials that travel to the soma where they are integrated. [modified from [13]]*

the *in vivo* recordings and showed that the spatial extent of stimuli determines the firing behaviour of sensory neurons in the electrosensory system demonstrating how delayed inhibitory feedback enables neurons to toggle between two distinct firing states, with each state connected to a behaviorally relevant stimulus [46].

In the human brain, almost every neuron could burst if it was pharmacologically manipulated (forced bursting), however, some neurons in certain areas of the human brain burst autonomously due to the somatic-dendritic interplay [41]. In some neurons, a somatic spike can excite the dendritic tree of the soma resulting in a delayed spike there that will depolarize the soma generating another somatic spike which again excites the dendritic tree and so fourth (see Figure 4.3).

In the CA1 region of the hippocampus, more than 80% of the neurons are non bursting neurons, whereas the remaining ones exhibit some form of bursting activity due to the interplay of fast ionic membrane currents that cause spiking activity and slower ionic membrane currents modulating this activity (intrinsic bursting) [41]. While fast ionic membrane currents

4.1. Physiological Mechanisms of Bursting

are the reason for the generation of AP's, they are unable to modulate the parameters, e.g. the spiking frequency, that characterize the spiking pattern because the generation of an AP is an all-or-nothing event, i.e. the underlying mechanism which leads to the generation of an AP is always the same. Only the addition of slow intrinsic membrane currents which typically build up during continuous firing activity will eventually result in the termination of the spike train due to hyperpolarization. In this way different spiking patterns can be generated by varying the *interburst frequency*, i.e. the period of quiescence between two consecutive bursts, or the *intraburst frequency*, i.e. the period of quiescence between two consecutive spikes within a burst. Experimental research aimed at identifying the slow membrane currents suggest that in order to produce the necessary inhibitory effect that terminates the spike train, the neuron must possess at least one of four known mechanisms: The neuron must either activate an outward current, e.g. a K^+-current or deactivate an inward current, e.g. a Ca^{2+}-current whereas this activation/deactivation can be either voltage-gated or Ca^{2+}-gated [41]. All four mechanisms have been documented *in vitro* [52–55].

Hippocampal pyramidal neurons can be categorized into five different classes, as illustrated in Figure 4.4: (1) non-bursting neurons (NB) generate trains of single spikes in response to depolarizing pulses of direct current and single spikes in response to a brief superthreshold pulse of current; (2) high threshold bursters (HTB) generate bursts in response to strong long pulses of direct current but fire single spikes in response to weaker or brief pulses of current; (3) grade I low threshold bursters (LTB I) generate bursts in response to long pulses of direct current but single spikes in response to brief pulses of currents; (4) grade II low threshold bursters (LTB II) generate bursts in response to brief pulses and (5) grade III low threshold bursters (LTB III) generate rhythmic bursts spontaneously, i.e. independent on the neuron's input. This classification does not imply any fundamental differences in the ionic mechanisms underlying the generation of spikes but a quantitative one, since pharmacological manipulations can gradually and reversibly transform a NB neuron into a LTB III neuron [56–60]. The behaviour of LTB II and LTB III neurons is of special interest for the concept of this thesis due to their properties of generating bursts in response to brief pulses of current. These neurons possess two populations of voltage-gated Na^+-channels with different threshold values, whereas one population facilitates fast ionic membrane currents which are responsible for the generation of spikes while the other population facilitates slow ionic membrane currents which are responsible for the spiking modulation and keep the burst alive. The current carried by ions passing through the slow channels is < 5% of the current carried by ions passing through the fast channels. The threshold value for the fast channels is slightly higher (about 10 mV [61]) than the value for the slow channels. Consequently, the slow channels are activated before the fast channels in response to depolarizing input. A depolarizing input current, which activates the slow channels, triggers a regenerative, positive feedback cycle which depolarizes the membrane potential further and further until the threshold value

Chapter 4: Bursting Neurons

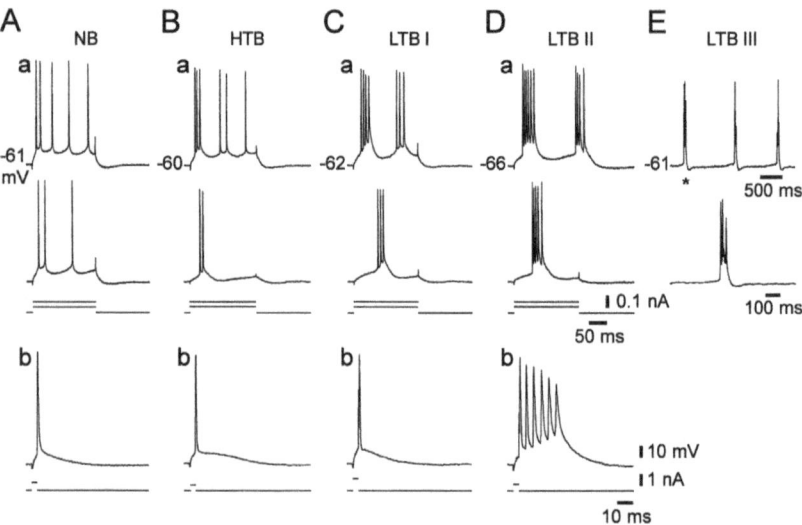

Figure 4.4.: *Hippocampal pyramidal neurons can be categorized into five different classes: (1) non-bursting neurons (NB) generate trains of single spikes in response to depolarizing pulses of direct current and single spikes in response to a brief superthreshold pulse of current; (2) high threshold bursters (HTB) generate bursts in response to strong long pulses of direct current but fire single spikes in response to weaker or brief pulses of current; (3) grade I low threshold bursters (LTB I) generate bursts in response to long pulses of direct current but single spikes in response to brief pulses of currents; (4) grade II low threshold bursters (LTB II) generate bursts in response to brief pulses and (5) grade III low threshold bursters (LTB III) generate rhythmic bursts spontaneously, i.e. independent on the neuron's input. [taken from [56]]*

for the fast channels is reached and the neuron generates a spike. After the first spike, the membrane potential is repolarized through the opening of voltage-gated K^+-channels and drops below the threshold value for the fast channels but stays above the threshold value for the slow channels so that the triggered positive feedback cycle stays active, allowing the generation of another spike. In this way, several consecutive spikes are generated until the slow channels start to close so that the membrane potential drops below the threshold value for the slow channels which terminates the spike train. Note that the time during which the

4.1. Physiological Mechanisms of Bursting

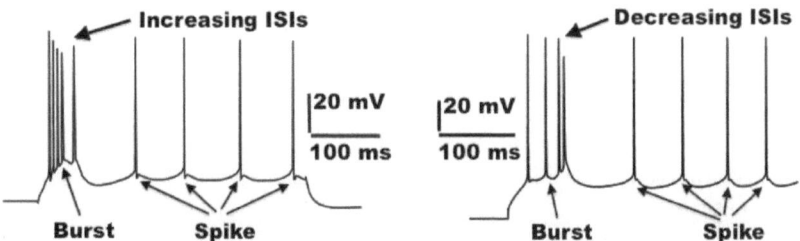

Figure 4.5.: *Bursting can be with increasing or decreasing interspike intervals (ISIs). Both traces show bursting activity followed by regular spiking activity.* **left:** *intrinsically bursting neuron of a cat;* **right:** *pyramidal neuron of rat visual cortex. [modified from [56]]*

slow channels are opened, the fast channels open and close several times while each opening and closing cycle generates a spike, eventually leading to the generation of a burst. The ISIs of bursts caused by fast and slow ionic membrane currents are increasing before the spike train is terminated, whereas the ISIs of burst as a result of the somatic-dendritic interplay are typically decreasing up to the point when a somatic spike falls into the refractory period of the dendrites and thus, fails to evoke dendritic response which stops the burst [62], as illustrated in Figure 4.5.

Info-box:
The neocortex is part of the cerebral cortex, a sheet of neural tissue that is the outermost to the cerebrum of the mammalian brain. It plays an important role for higher brain functions such as sensory perception, generation of motor commands, spatial reasoning, consciousness and language and is made up of six layers, labelled I to VI with I being the outermost and VI being the innermost, whereas each layer is different in terms of neurons and connectivity. The Pre-Boetzinger Complex is a cluster of interneurons in the ventrolateral medulla of the brainstem, which is essential to the generation of respiratory rhythm in mammals. The medulla oblongata, the lower half of the brainstem, contains the cardiac, respiratory, vomiting and vasomotor centers and deals with basic autonomic functions such as breathing, heart rate and blood pressure. [65, 66]

In summary, three different mechanisms underlying neuronal bursting activity have been explained: (1) Stimulus induced oscillation produced by inhibitory feedback, (2) the somatic-dendritic interplay with typically decreasing ISIs and (3) the interplay of fast and slow ionic membrane currents with typically increasing ISIs. Bursting neurons have been found in different structures of the brain, e.g. intrinsically bursting pyramidal neurons in the neocortical layer V [63], pyramidal chattering neurons in the neocortical layers II-IV (mainly layer III) [64], low-threshold pyramidal neurons in the CA1 region of the hippocampus [58] or rhythmic bursting neurons in the pre-Boetzinger complex that control the respiration cycle [65]. The above examples give reason to believe that the physiological distinctions of neurons in the brain, especially in the neocortex, have morphological correlates [63]. **Furthermore, since bursting has been found if many different structures of the brain, one key question poses itself: What is so special about bursting as a type of neuronal information? In the next section, several hypotheses try to give a possible answer to this question.**

4.2. Bursts as a Unit of Neuronal Information

Many hypotheses on the importance of bursting activity in neural computation have been proposed:

- **Bursts are more reliable than single spikes** in evoking responses in postsynaptic neurons, because EPSPs from each spike within a burst may add up resulting in a superthreshold EPSP [41].

- **Bursts overcome synaptic transmission failure.** It has been suggested that postsynaptic response of cells, although sensitive to single spikes, may fail because the release of neurotransmitter in response to a single PSAP does not occur, however, a bombardment of PSAPs, as it can be found in a burst, makes the release of neurotransmitter more likely [43].

- **Bursts facilitate transmitter release**, whereas single spikes do not [43]. Some synapses exhibiting strong STF are insensitive to single spikes or even short bursts and therefore transmitter release can only be facilitated by long bursts where the last spikes within the burst have stronger effect than the previous spikes.

- **Bursts evoke long-term potentiation** and may affect synaptic plasticity much more than single spikes [43, 44]. In some cases single spikes can only influence short-term plasticity which at some point recovers to its initial state, meaning that no long-term change in plasticity takes place.

- **Bursts have a higher signal-to-noise ratio than single spikes** [67]. In some neurons, the burst threshold is higher than the spike threshold so that it is more likely for a spike to be generated as the result of input noise, whereas the generation of bursts requires stronger input making them more insensitive to noise.

- **Bursts can be used for selective communication.** Some neurons have subthreshold membrane potential oscillation instead of a constant resting potential and as postsynaptic cells they are sensitive to the frequency content of the input. Depending on the intraburst frequency, some bursts resonate with the oscillation and elicit postsynaptic response, others do not [42].

- **Bursts can resonate with short-term plasticity** making a synapse a band-pass filter [42]. Some synapses that exhibit STF and STD are most sensitive to bursts with a certain resonant intraburst frequency. Such bursts evoke strong facilitation but only weak depression so that the effect on the postsynaptic neuron is maximized.

- **Bursts encode different features** of sensory input than single spikes [68, 69]. Neurons in the electrosensory lateral line lobe (ELL) of weakly electrical fish are able to distinguish between communication signals, processed as bursts, and prey signals, processed as single spikes (see section 4.1) [46]. Pyramidal neurons in the thalamus of the visual system compose natural scenery by encoding stimuli which inhibit the neuron with bursts for a period of time and then rapidly excite it upon spike-train termination [70].

- **Bursts have more informational content than single spikes** when analyzed as unitary events [71]. Additional informational content is founded in the burst duration or the fine temporal structure of interburst and intraburst intervals.

- **Bursts are necessary to drive synaptic refinement**, e.g. in the lateral geniculate nucleus (LGN), the primary relay center for visual information received from the retina [45]. A burst-based hebbian learning rule, burst-timing-dependent plasticity (BTDP), is found to be necessary for the development of the retina during the period of eye segregation.

The previous hypotheses can be summarized with the conclusion that burst input, in contrast to single spike input, is more likely to have a stronger impact on postsynaptic cells for various reasons. Some even believe that bursts are all-or-nothing events, whereas single spikes may be noise [41]. **Two of the above hypotheses already mentioned the selective communication via burst resonance as an example for the functional significance of bursting in the nervous system which is explained in detail in the next section.**

4.3. Bursting for Selective Communication

Complementary to the dominant hypotheses that bursts are needed to increase the reliability of communication between neurons [43], an alternative hypothesis has been proposed that is: bursts with specific resonant interspike frequencies are more likely to cause a postsynaptic neuron to fire than bursts with higher or lower frequencies [42]. This frequency dependence occurs either at the level of individual synapses due to the synapse's interplay between STF

Chapter 4: Bursting Neurons

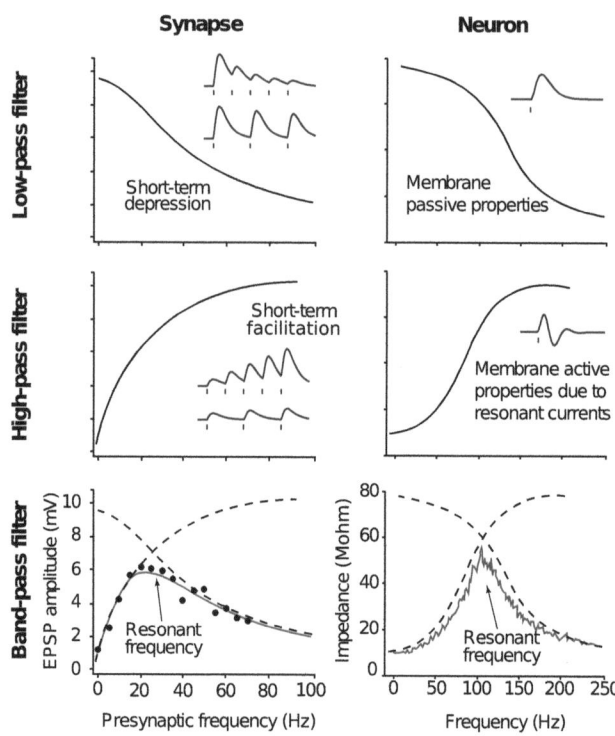

Figure 4.6.: Bursts can be used for selective communication via resonance. **left:** Resonance at the synaptic level due the interplay of STF and STD. Synapses exhibiting STD act as a low-pass filter, whereas synapses exhibiting STF act as a high-pass filter. The combination of both competing effects at a single synapse allows it to act as a band-pass filter with a resonant frequency for which the synapses EPSP amplitude is maximal. The insets show the corresponding PSPs for high and low frequency inputs. **right:** Resonance at the postsynaptic cellular level due to subthreshold membrane potential oscillation. The interplay of a neuron's passive and active membrane properties, each with a different frequency dependence, leads to a resonant frequency for which a postsynaptic response is most likely. The insets show the time course of typical PSPs. [modified from [42]]

4.3. Bursting for Selective Communication

and STD or at the postsynaptic cell level due to the cell's subthreshold membrane potential oscillation and resonance. In some synapses the competing effects of STF and STD have different dependencies on the input frequency [72–75], which causes such synapses to act like a band-pass filter, i.e. there is a specific resonant interspike frequency for the synaptic input which has maximum effect on the postsynaptic neuron, as illustrated in Figure 4.6. Given that there are many synapses with different resonant frequencies belonging to the same presynaptic neuron [72, 73] such synaptic filtering provides a potent tool for selective communication between neurons. By changing the intraburst frequency, a presynaptic neuron can selectively affect some postsynaptic neurons but not others [42].

Some neurons exhibit subthreshold membrane potential oscillation (sometimes referred to as being a resonator [76]) which are often caused by slow intrinsic membrane currents. The interplay of such a neuron's passive properties, i.e. the intrinsic dependence of V_m on PSPs, and active properties, i.e. the alternating activation of persistent low threshold Na^+- and K^+-currents that are responsible for the membrane potential oscillation, have different dependencies on the input frequency, causing the neuron's response to be sensitive to the precise timing of input pulses, as illustrated in Figure 4.6. The effect on such neurons is maximal when the intraburst frequency of the neurons's input is close to the neuron's natural frequency, i.e. their subthreshold membrane potential oscillation frequency which depends on the temporal sequence of the opening and closing mechanisms of the specific ion channels that are responsible for these oscillation. [42]. This is because input spikes arriving with the right frequency, i.e. arriving at a point in time when the postsynaptic membrane potential is at a maximum, add up causing response augmentation, whereas a frequency mismatch leads to response attenuation, as illustrated in the top part of Figure 4.7. The bottom part of Figure 4.7 illustrates the same idea using a V-I phase diagram. Its dynamics can be depicted as trajectories in the V-I phase plane indicated by the black and red arrows. An incoming pulse displaces the trajectory from its stable equilibrium state (central black circle) leading to damped oscillation (black arrow). The effect of the second pulse depends in the temporal structure of the doublet (a burst consisting of two spikes). If the ISI is close to the interval between two maxima of the membrane oscillation, the trajectory will be pushed away from the equilibrium, thereby increasing the amplitude of the oscillation, whereas a ISI different from the interval between two maxima of the membrane oscillation will cause the trajectory to be pushed back towards the equilibrium, canceling the effect of the first pulse. Both mechanisms explained above, resonance at the synaptic level and resonance at the postsynaptic cellular level, evidently show that a presynaptic neuron can selectively communicate with some postsynaptic neurons but not with others simply by varying its intraburst frequency because a burst with a specific intraburst frequency could resonate with some synapses or postsynaptic neurons but not with others, depending on their natural resonance frequency. In this way, bursting neurons have the ability to switch between postsynaptic neurons on the

Chapter 4: Bursting Neurons

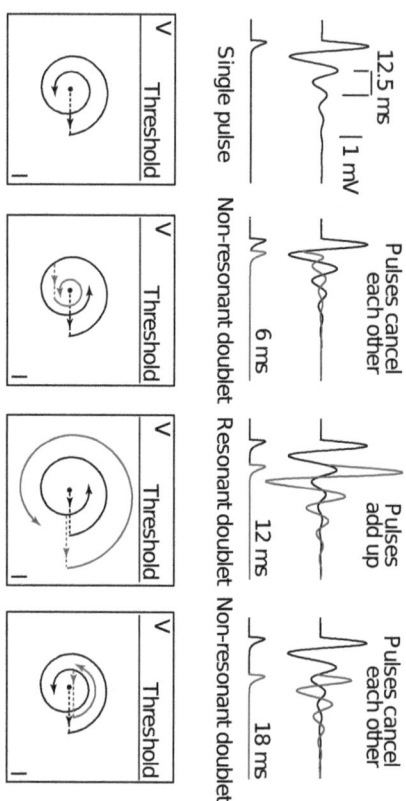

Figure 4.7.: A neuron's subthreshold membrane potential oscillation lead to resonance and frequency preference. **top:** Simulation of a neuron exhibiting subthreshold oscillation with a frequency of 80 Hz in response to synaptic stimuli. Pulses arriving with an input frequency close to the neuron's oscillation frequency add up which eventually results in a superthreshold excitation, whereas pulses arriving with a frequency mismatch cancel each other out. **bottom:** Corresponding V-I phase portraits. The black and red arrows, representing the first and second input pulse respectively, illustrate the temporal change in membrane potential depending on the input frequency. [modified from [42]]

time scale of tens of millisecond without affecting long-term synaptic modifications [42].

Bursting for selective communication is just one example out of some known and presumably many unknown other cases which shows that bursting plays an important role for complex neural computation in several areas of the brain, mainly because of the many additional possibilities to encode neural information resulting in complex functioning of equally complex neural networks containing versatile single components, neurons and synapses, which offer various ways to take advantage of processing bursts as a unit of neuronal information. **The importance of bursting for neural computation demands a complete, yet simple mathematical description of this activity in order to provide the possibility to model the dynamics of bursting neurons within complete neural networks.**

4.4. Modeling Neuronal Bursting Activity

There are three mathematical models, each of which having advantages and flaws, that are commonly used by neuroscientists to model neuronal activity: (1) the Integrate-and-Fire (I&F) model, (2) the Resonate-and-Fire (R&F) model and (3) the Quadratic Integrate-and-Fire (QI&F) model. In order to take the best of each of the three models, Eugene Izhikevich, a neuroscientist with main research interests in nonlinear dynamical systems, spiking neural networks and large-scale simulations of the brain, combined all three models into one single simple model [56]. This model has only four dimensionless parameters which influence how modeled neuronal activity looks like. **Understanding in which way varying these parameters influences neuronal activity helps to create a physiological analogy and, with respect to the concept of an artificial PCM bursting neuron, a physical analogy to these parameters**

4.4.1. The Integrate-and-Fire Model

The leaky Integrate-and-Fire (I&F) model is an idealization of a neuron having ohmic leakage current, which sustains the resting potential value, and a number of voltage-gated currents that are completely deactivated at rest. The subthreshold behaviour of such a simplified neuron can be described by the linear differential equation

$$C_m \cdot \dot{V}_m = I_{app} - G_{leak}(V_m - E_{leak}),$$

where $C_m = A \cdot c_m$ is the total membrane capacitance with A being the total membrane area, V_m is the neuron's membrane potential, $G_{leak} = A \cdot g_L = A/r_m$ is the total membrane conductance, E_{leak} is the neuron's leaking potential, i.e. resting potential value, and I_{app} is an externally applied inward current, i.e. an EPSC or an IPSC depending on its sign. The term $G_{leak}(V_m - E_{leak})$ describes the current flow per unit area out of the neuron through "leak"

Chapter 4: Bursting Neurons

channels, i.e. resting channels, in the neuron's membrane. When V_m reaches the threshold value E_{th}, the voltage-gated currents instantaneously activate and the neuron is said to fire a spike after which V_m is reset to a value E_K below E_{leak} in order to model hyperpolarization. After rescaling, the model can be rewritten as:

$$\dot{v} = b - v$$

and if $v = 1$ then $v \leftarrow 0$, where $v = b$ is the resting state, $v = 1$ is the threshold value and $v = 0$ is the reset value. Important properties of this model are: (1) all spikes are assumed to be of equal size and duration, (2) a well defined threshold value exists and is the only condition for spiking, (3) a relative refractory period is modeled for $E_K < E_{leak}$ during which the neuron is less excitable, (4) the difference between excitatory input ($I_{app} > 0$) and inhibitory input ($I_{app} < 0$) is considered and (5) the neuron encodes input strength into spiking frequency. The main flaw of this model is that the actual generation of spikes is not considered by its mathematical equation and modeled neurons are only said to spike upon reaching E_{th}. Spikes that are often depicted in connection with this model are added by hand in order to illustrate the modeled neuronal activity (see Figure 4.8). [56]

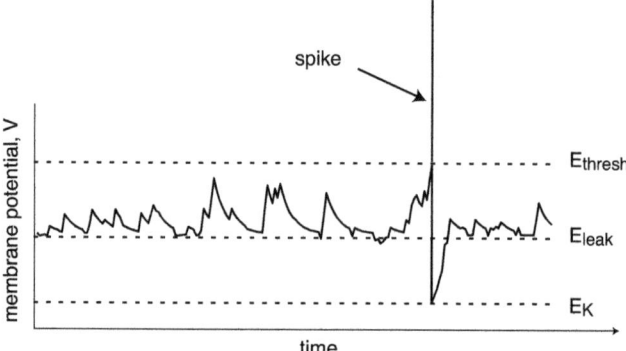

Figure 4.8.: Neuronal activity modeled with the leaky Integrate-and-Fire model. The trace includes noisy input and illustrates spiking behaviour upon reaching the threshold value after which the potential is reset to model hyperpolarization. The spike is manually added for aesthetic purposes and is no result of the model itself. [taken from [56]]

4.4.2. The Resonate-and-Fire Model

The Resonate-and-Fire (R&F) model is a two dimensional extension of the I&F model incorporating an additional low-threshold persistent K$^+$-current that is partially activated at rest and is responsible for a neuron's subthreshold membrane potential oscillations. This resonating current's magnitude is denoted by an additional parameter W_m in the equation of the I&F model:

$$C_m \cdot \dot{V}_m = I_{app} - G_{leak}(V_m - E_{leak}) - W_m$$

with

$$\dot{W}_m = (V_m - V_{1/2})/k - W.$$

If $V_m \geq E_{th}$ then $V_m \leftarrow E_K$, similar to the I&F model and in addition $W_m \leftarrow W_K$, where W_K is some parameter. When the resting state is a stable focus, the model can be rewritten in complex coordinates:

$$\dot{z} = (b + i\omega) \cdot z + I,$$

where $b + i\omega \in \mathbb{C}$ is the complex eigenvalue of the resting state and $z = x + iy \in \mathbb{C}$ is the complex variable describing the damped membrane potential oscillations with the frequency ω around the resting potential value. The real part x, a current-like variable, describes the dynamics of the resonant current and PSCs, whereas the imaginary part y, a voltage-like variable, describes the neuron's membrane potential. The neuron is said to fire a spike when y reaches the threshold $y_{th} = 1$ which is therefore a horizontal line in the complex plane that passes through $i \in \mathbb{C}$. After firing a spike, z is reset to z_{reset}. The dynamics of the R&F model explained above are illustrated in Figure 4.9. The R&F model illustrates the most important features of resonators which are amongst others: damped subthreshold membrane potential oscillations and input frequency preference (see 4.3). The main flaw of this model is the same as for the I&F model that is: the actual generation of spikes is not considered and must be added by hand. [56]

4.4.3. The Quadratic Integrate-and-Fire model

The Quadratic Integrate-and-Fire model eliminates the main flaw of the I&F and R&F model. The rescaled equations of the I&F model are modified by replacing $-v$ with v^2 in order to consider the actual generation of a spike:

$$\dot{v} = b + v^2$$

and if $v = v_{peak}$, then $v \leftarrow v_{reset}$. Here, v_{peak} is not a threshold but the potential value at which the a spike is cut off, i.e. the potential actually rises and is cut off so that an actual potential peak is generated. Although this model can be best understood by a mathematical

Chapter 4: Bursting Neurons

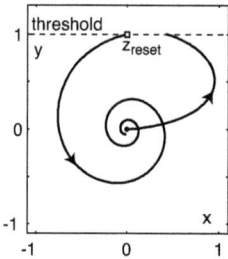
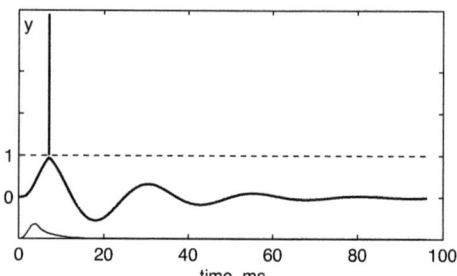

Figure 4.9.: *Neuronal activity modeled with the Resonate-and-Fire model.* **left:** *The complex plane illustrating the subthreshold dynamics of x and y and therefore analogously the V-I characteristics (see also the bottom part of Figure 4.7).* **right:** *simulated neuronal activity with a single spike and subsequent subthreshold membrane potential oscillations. Simulation parameters: $b = -0.05$, $\omega = 0.25$ and $z_{reset} = i$. The spike was added manually by hand. [taken from [56]]*

description on grounds of bifurcation, an elaboration on this topic is left out because the exact mathematical understanding is not important for the concept of this thesis and can be found elsewhere, e.g. in [56]. [56] For further reading, it is only important to keep in mind, that the QI&F model, unlike its predecessors, is a genuine integrator and is therefore much closer to biological neuronal spiking behaviour than the I&F and the R&F model.

4.4.4. The Simple Model of Choice

Izhikevich's simple model combines the advantageous properties of all three models. Its dynamics is described by the following equations:

$$\dot{v} = I + v^2 - u$$

and

$$\dot{u} = a \cdot (b \cdot v - u),$$

and if $v \geq 1$ then $v \leftarrow c$ and $u \leftarrow u + d$. In the above equations, v is the neuron's membrane potential and u is the recovery current (in analogy to the parameter W of the R&F model), i.e. the sum of all slow currents that modulate the spike generation mechanism. In addition, the simple model has four dimensionless parameters (a, b, c and d) that influence the simulated neuronal activity. a is the recovery time constant and determines how strong the slow currents

4.4. Modeling Neuronal Bursting Activity

affect the spiking generation mechanisms. The sign of b determines whether u is amplifying ($b < 0$ or a resonant variable ($b > 0$). c is the voltage reset value after a spike is generated and d is the total amount of outward minus inward currents which are activated during the spike generation. Figure 4.10 illustrates how neuronal activity, or more precisely the generation of spikes, can be modeled with the simple model. Depending on the values of a and b, the

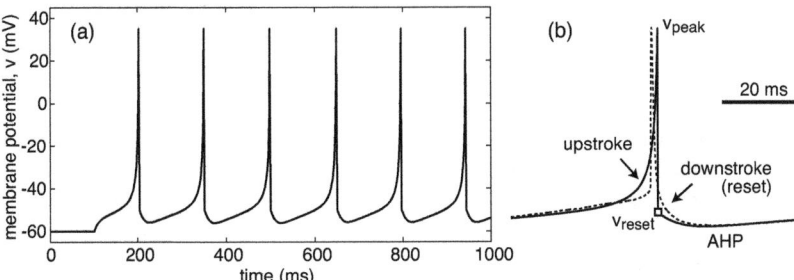

Figure 4.10.: *Neuronal activity modeled with the simple model. **a)** Output of the MAT-LAB code (which is freely accessible in [56]) simulating the simple model. **b)** Comparison of a simulated trace (solid curve) and experimental trace (dashed curve) shows major discrepancies that are marked by arrows. In contrast to the I&F model, the simple model (right) actually generates spikes at a soft dynamic threshold range that are cut off after which the potential is reset. [taken from [56]]*

modeled neuron can be an integrator or a resonator. The parameters c and d on the other hand do not affect the steady-state subthreshold behaviour but instead, take into account the activation of high threshold voltage-gated currents that are activated during the spike affecting only the after-spike transient behaviour. By varying these four parameters, basically every kind of neuronal activity can be modeled, as shown in [56], however, only bursting neuronal activity is of increased interest for the concept of this thesis. Hence, the influences of the parameters on neuronal bursting activity, i.e. on the two basic characteristics of a burst (ISI and burst duration) were analyzed. The recovery constant a mostly regulates the pace of the modeled system, i.e. it determines the speed with which the system recovers after generating a spike and therefore strongly influences the ISIs within bursts, while there was no influence on the number of spikes within a burst observed, i.e. the burst duration changes accordingly to the ISIs. The voltage reset value c determines the modeled system's ability to generate spikes during a single burst, i.e. it models the strength with which the system has to be excited in order to generate spikes. d determines how much ability of the system

Chapter 4: Bursting Neurons

to spike is consumed for each spikes, i.e. it determines how many spikes can be generated within a burst. The effects of b on the characteristics of a burst are more complex and not really relevant considering only the the basic characteristics that were analyzed.

In summary, the relation between how much spiking ability is consumed by the generation of spikes, determined by d, and how much of this ability is available for a single burst, determined by c which is coupled to a recovery phase, is decisive. Moreover, the recovery constant a generally determines the over all pace of the modeled system.

4.5. An Overall View

Neuronal bursting activity occurs at several different regions in the brain and presumably plays an important role in neuronal communication. Three different mechanisms have been documented, each of which is fundamentally different in generating bursting behaviour. In comparison to single spiking activity, bursts as a unit of neuronal information provide several additional features that lead to additional complexity in neuronal communication. A simple model introduced by Izhikevich which describes neuronal bursting activity has four dimensionless parameters with three of them having specific physiological analogies that can be later connected to physical analogies considering the concept of an artificial PCM bursting neuron. **The next chapter introduces a concept for an electronic device capable of exhibiting neuronal bursting behaviour which could serve in an experiment complementing computational studies on neuronal bursting behaviour.**

CHAPTER 5

A PCM Bursting Neuron

A today's common CMOS neuron with only one single synaptic input node contains at least 58 transistors, whereas each additional input node would require another ten transistors. Furthermore, such a neuron contains only one single threshold circuit which makes the neuron's output behaviour independent on its level of activation. In order to provide this functionality, additional threshold circuits, each consisting of eight transistors, have to be added [77]. Consequently, artificial neuronal behaviour emulated on the basis of CMOS technology is the result of a huge number of transistors forming several electrical circuits that can be used for very large scale implementations (VLSIs) of pulsed neural networks. Although these neurons provide the possibility to emulate the integrative nature of biological neurons, they consume a huge amount of space and considering the barriers that are posed to downscaling by the nano era (see chapter 1) they barely consider the future problems of computational progress. Even more importantly, the emulated behaviour is constructed in a highly artificial fashion and is in now way founded in any inherent physical behaviour that can be awarded to a specific component. Furthermore, the author has no knowledge about any experimental studies with respect to neuronal bursting behaviour, although it is supposed to play an important role in the communication between neurons in the brain (see chapter 4). **Because of the assumed importance of neuronal bursting, this chapter introduces the first thoughts on a path to an artificial phase-change bursting neuron based on the reversibility of the switching behaviour between the a-off and a-on state which allows to induce voltage-controlled relaxation oscillation in phase-change cells [78] and might be used to emulate neuronal bursting activity in an experimental way to complement computational studies.**

5.1. Voltage-Controlled Relaxation Oscillation in a PCM Device

In 2008, Ielmini et al. showed that the NDR of the amorphous state in the I-V characteristics in combination with a parallel capacitive element can be exploited to generate periodic relaxation oscillation between the a-off and a-on state [78]. Figure 5.1 shows the measured I-V characteristics of the amorphous state for a μ-trench GeSbTe-based cell at which the experiments were conducted on. The characteristics displays the typical high resistance a-off

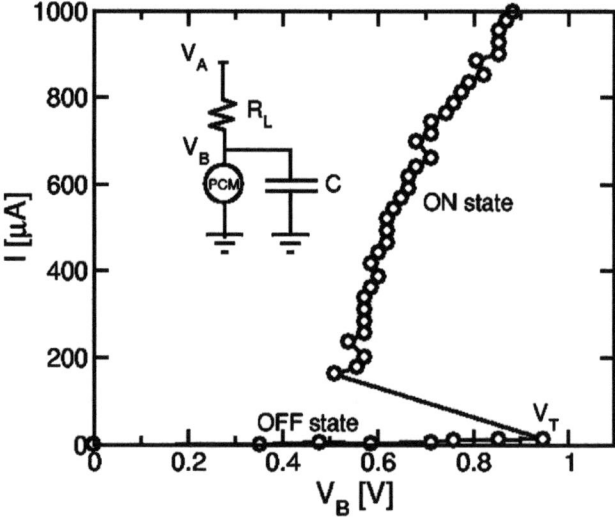

Figure 5.1.: *Measured I-V characteristics for a the μ-trench PCM cell in the amorphous state. The I-V curve displays a clear NDR characteristic which is the basis of the voltage oscillation. The inset shows a schematic of the circuit under test. [taken from [78]]*

state ($R_{off} > 1\,M\Omega$) and the threshold switching behaviour which appears as a dramatic voltage snapback at the threshold voltage $V_T \approx 0.95\,V$. After the threshold switch, the current rises in the low resistance a-on state ($R_{on} \approx 1\,k\Omega$). The I-V characteristics was measured by applying electrical pulses to a PCM cell in series with a load resistance of $R_L = 3.3\,k\Omega$ influencing the cell current as follows:

$$I = \frac{V_A - V_B}{R_L},$$

5.1. Voltage-Controlled Relaxation Oscillation in a PCM Device

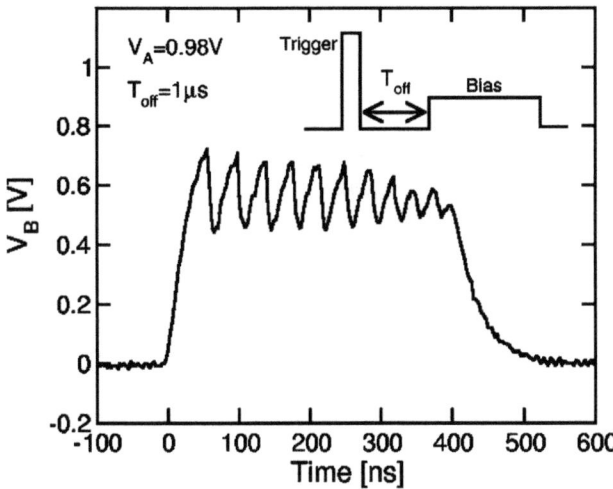

Figure 5.2.: Measured voltage oscillation across the PCM cell V_B as a function of time. The inset shows the pulse sequence of the applied voltage V_A corresponding to the oscillation which consists of a trigger pulse to reset the cell, a variable sleep time T_{off} and the bias pulse. [taken from [78]]

where V_A is the applied voltage and V_B is the voltage across the PCM cell. The inset shows the complete circuit used to induce the oscillation behaviour with the PCM cell itself and a parallel capacitance of $C = 12\,\text{pF}$ due to the parasitic influences which generally results from the cell's geometry and other factors, e.g. the measurement tip, the contact pad, etc. The parasitic inductance in the circuit is smaller than 1 nH and therefore negligible. Biasing the electrical circuit appropriately leads to voltage oscillation across the PCM cell as a result of the NDR. Figure 5.2 shows typical oscillation behaviour of V_B while a two-pulse signal was applied containing an initial high current trigger pulse which resets the PCM cell. After a sleep time T_{off}, there follows a second bias pulse, slightly above V_T, during which voltage oscillations across the cell are observed. For a time $T_{off} = 1\,\mu s$ after the trigger pulse, a bias pulse of 400 ns width at $V_A = 0.98\,\text{V}$ was applied leading to a damped sawtooth waveform featuring eleven spikes as a result of repetitive switching/relaxation events. The oscillation frequency f_{osc} can be simply tuned by controlling the bias voltage. Figure 5.3 shows f_{osc} as a function of V_A during the bias stage, for two different sleep times $T_{off} = 0\,\mu s$ and $1\,\mu s$. f_{osc} was calculated from the peak in the fast Fourier transform (FFT) of the measured voltage

Chapter 5: A PCM Bursting Neuron

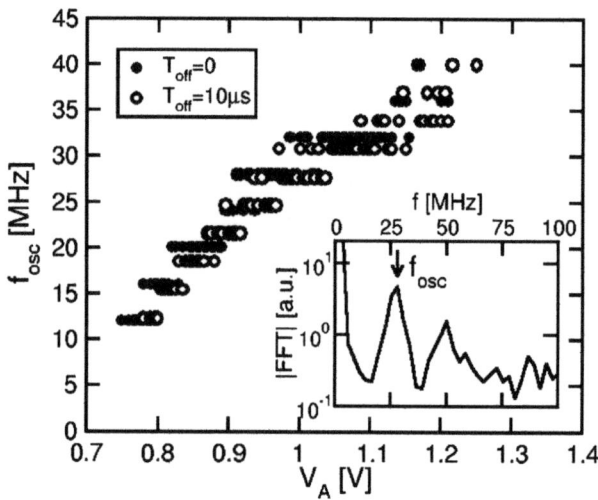

Figure 5.3.: *Measured oscillation frequency f_{osc} for different sleep times T_{off} as a function of the applied voltage V_A. f_{osc} can be controlled by varying V_A whereas the influence of T_{off} is negligible. The inset shows the measured FFT of the signal in Figure 5.4 displaying the first harmonic peak at around 27 MHz. [taken from [78]]*

oscillation V_B as shown in the inset of Figure 5.3 and is practically independent on T_{off}.

The voltage oscillations are explained by the switching and recovery dynamics in the PCM. In the a-off state, the material switches into the a-on state at a threshold voltage V_T and the current suddenly rises dramatically at almost constant voltage because the characteristic time for capacitance discharge $\tau_C = R_{ON} \cdot C \approx 10\,\text{ns}$ is long compared to the switching time $\tau_s \approx 0.1\,\text{ns}$, i.e. the actual time which passes during the switch from the a-off state into the a-on state τ_s is short compared to the time which passes until the electrical circuit reacts to the change in resistance of the PCM cell. After the PCM switched into the low resistance a-on state, the voltage across the PCM cell drops and the voltage across the load resistance R_L rises according to Kirchhoff's circuit laws. This change in voltages across the devices does not happen instantly but in a rather continuous fashion due to capacitance discharge. In this way, the voltage across the PCM cell relaxes until it drops below the holding voltage V_H, the minimum voltage which has to be applied to the PCM cell in order to sustain the a-on state [79]. At voltages below V_H, the PCM's a-off state is continuously recovered. This

5.1. Voltage-Controlled Relaxation Oscillation in a PCM Device

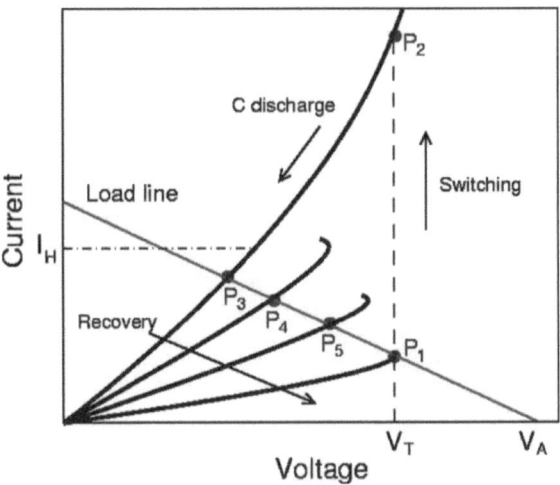

Figure 5.4.: Schematic illustration of the oscillation mechanism in the PCM cell. Starting at a subthreshold point, reaching V_T at P_1 leads to a current rise to P_2 upon switching into the a-on state. The current decreases to a point P_3 below the the holding voltage V_H and accordingly below the holding current I_H as the capacitance discharges, and then the voltage increases as the a-off state continuously recovers thought the points P_4 and P_4. Switching occurs again once V_T is reached again. [taken from [78]]

transition is driven by two factors: (1) the recovery transient ($\tau_r \approx 10\,\text{ns}$) which is needed to restore thermal and electronic equilibrium within the PCM [80] and (2) the capacitance charging time $\tau_C = R_L \cdot C \approx 40\,\text{ns}$. After the capacitance is charged up to the point when V_T is reached, another switching event and another oscillation cycle is induced. This accounts for the magnitude of the oscillation frequency $f_{osc} \approx (R_L \cdot C)^{-1} \approx 25\,\text{MHz}$ which can be controlled by varying R_L, C and V_A. An increase of V_A leads to an increase in f_{osc} due to a faster exponential rise in V_B, which is connected to the capacitance charge, leading to a shorter time to reach V_T again. Damping of the oscillation behaviour might be explained by a progressive decrease of the amorphous volume in the PCM layer as a result of repeated cycling. At Each switching transition, a part of the programmable PCM region is amorphized by fast melting (from P_1 to P_2) and subsequent quenching (from P_2 to P_3) [81]. The volume of this region depends on the current flow through the PCM device at P_2 ($I_{P_2}(V_T)$) which is an increasing function of V_T. On the other hand, the amorphous volume controls V_T due to the relationship between the amorphous PCM thickness and V_T [38, 82]. Thus, if switching occurs at a voltage V_{T1} resulting in an amorphous region that is slightly smaller than the original one, a subsequent switch will occur at a voltage V_{T2} which is smaller than V_{T1}. In this way, V_T progressively collapses during repeated oscillation spikes resulting in damped oscillation behaviour. In addition, heating induced crystallization during the transition from P_2 to P_3 may also contribute to a decrease of the amorphous volume and hence, of V_T. Figure 5.4 illustrates the oscillation mechanism explained above. **In order to emulate neuronal bursting activity through voltage-controlled relaxation oscillations, it is necessary to examine how the physical parameters of the system influence the oscillation behaviour so that it can be manipulated to fit neuronal bursting behaviour.**

5.2. The Analogy to Hippocampal Pyramidal Bursting Neurons

Voltage-controlled relaxation oscillations can be observed as a result of the sudden decrease of measured voltage across the device under test (DUT), i.e. the PCM device. According to Kirchhoff's circuit laws, a second device, preferably an ohmic resistance, is required to be connected in series with the PCM device in order to control the exact voltage drops that have to be measured in case of oscillating behaviour. The resistance value for the load resistance R_L that has to be chosen in order to facilitate oscillation behaviour depends on the resistances of the DUT in the a-off and a-on state $\left(R_{DUT}^{off,on}\right)$ and can be calculated from

$$V_B^{off,on} = \frac{R_{DUT}^{off,on}}{R_L + R_{DUT}^{off,on}} \cdot V_A, \qquad (5.1)$$

when considering two conditions. (1) the maximum voltage across the PCM device has to be slightly higher than V_T in order to facilitate the transition from the a-off state into the a-on state, while (2) the minimum voltage across the PCM device has to be slightly lower than V_H in order to facilitate relaxation back into the a-off state, i.e.:

$$V_T < \frac{R_{DUT}^{off}}{R_L + R_{DUT}^{off}} \cdot V_A \qquad (5.2)$$

and

$$V_H > \frac{R_{DUT}^{on}}{R_L + R_{DUT}^{on}} \cdot V_A. \qquad (5.3)$$

A resistance range for R_L can be calculated by inserting (5.2) and (5.3) in (5.1):

$$R_{DUT}^{on} \cdot \left(\frac{V_A}{V_H} - 1\right) < R_L < R_{DUT}^{off} \cdot \left(\frac{V_A}{V_T} - 1\right) \qquad (5.4)$$

5.2. The Analogy to Hippocampal Pyramidal Bursting Neurons

In order to reduce or even avoid damping during oscillations caused by the progressive collapse of V_T as the result of variable a-on state currents and partial crystallization during repeated cycling, which would at some point terminate the artificial burst even though this might not be desired, one could use an ovonic threshold switching (OTS) material, named after S. Ovshinsky, with a low crystallization speed. A suitable candidate might be GeTe$_6$ which has a crystallization time of more than $100\,\mu s$, i.e. crystallization induced by nano second pulses is impossible, and its local amorphous region is not modified by different a-on state currents [79, 83]. For a GeTe$_6$ device fabricated in the classical bottom heater geometry [84] with a 20 nm thick GeTe$_6$ layer sandwiched between a TiN heater with 60 nm diameter, which is embedded in an isolating silicon nitride layer, and a 40 nm thick TiN top electrode, the electrical switching characteristics were examined [79]. The device's resistances in the a-off and a-on state were found to be $R_{DUT}^{off} = 80 \pm 10\,M\Omega$ and $R_{DUT}^{on} = 3 \pm 1\,k\Omega$, respectively, with $V_T = 1.60 \pm 0.02\,V$ and $V_H \approx 0.7\,V$. The resistance range for the load resistance, which has to be appropriately chosen in order to induce voltage oscillations in such a GeTe$_6$ device, can be calculated from (5.4):

$$\sim 4.285\,k\Omega < R_L <\sim 5\,M\Omega$$

with $V_A = 1.7\,V$. In order to emulate hippocampal pyramidal neuronal bursting behaviour, the parameters of the PCM bursting neuron must be carefully chosen so that the voltage oscillations fit neuronal bursting activity, thus, the amplitude and the temporal structure of the PCM bursting neuron must be tuned. The inverse of the relative amplitude A_{osc} of the voltage oscillations can be calculated from

$$A_{osc}^{-1} = \frac{V_B^{off}}{V_B^{off} - V_B^{on}}$$

and must be tuned in order to fit the relative amplitude of hippocampal pyramidal neuronal bursting behaviour which can be achieved by choosing the correct resistance ratio of R_L to R_{DUT}^{on}:

$$\begin{aligned} A_{osc}^{-1} &= \frac{V_B^{off}}{V_B^{off} - V_B^{on}} = \frac{\frac{R_{DUT}^{off}}{R_L + R_{DUT}^{off}}}{\frac{R_{DUT}^{off}}{R_L + R_{DUT}^{off}} - \frac{R_{DUT}^{on}}{R_L + R_{DUT}^{on}}} \\ &= \frac{R_{DUT}^{off}}{R_L + R_{DUT}^{off}} \cdot \left[\frac{R_{DUT}^{off}(R_L + R_{DUT}^{on}) - R_{DUT}^{on}(R_L + R_{DUT}^{off})}{(R_L + R_{DUT}^{off}) \cdot (R_L + R_{DUT}^{on})} \right]^{-1} \\ &= \frac{R_{DUT}^{off}(R_L + R_{DUT}^{on})}{R_L \left(R_{DUT}^{off} - R_{DUT}^{on} \right)} \approx 1 + \frac{R_{DUT}^{on}}{R_L}. \end{aligned}$$

Chapter 5: A PCM Bursting Neuron

Here, $R_{DUT}^{off} \gg R_{DUT}^{on}$ has been used in the last step. Thus, by choosing the correct value for R_L while having exact knowledge of R_{DUT}^{on}, it is possible to chose a relative oscillation amplitude which fits hippocampal pyramidal neuronal bursting activity. The exact value for R_L which is needed in order to emulate such bursting activity can be calculated after examining the relative amplitude of such neurons. The top part of figure 5.5 illustrates the varying relative amplitude of a hippocampal pyramidal LTB II neuron which can be calculated from the values d_1 and d_2:

$$A_{osc}^{-1} = \frac{d_2}{d_2 - d_1},$$

with $d_1 = |V_r - V_{min}|$, $d_2 = |V_r - V_{max}|$ and $V_r = -65\,\text{mV}$ [58]. An estimation for the first spike yields: $V_{min} = -52\,\text{mV}$ and $V_{max} = 23\,\text{mV}$, so that $A_{osc} \approx 1.173$. Thus,

$$1.173 \stackrel{!}{=} 1 + \frac{R_{DUT}^{on}}{R_L}$$

$$\Rightarrow R_L = \frac{R_{DUT}^{on}}{0.173} \approx 17.34\,\text{k}\Omega.$$

In order to reproduce voltage oscillations with the same relative amplitude as of a hippocampal pyramidal LTB II neuron, a load resistance value of $R_L \approx 17.34\,\text{k}\Omega$ has to be chosen. Additionally, the value of A_{osc}^{-1} needs to be gradually changed from $A_{osc}^{-1} = 1.137$ for the first spike to $A_{osc}^{-1} = 1.558$ for the last spike within the burst of Figure 5.5 to emulate the damped bursting behaviour of hippocampal pyramidal neurons. Damping of the amplitude during voltage oscillations can not be achieved by varying R_L when a common resistor is used since the variation of its resistance value must be possible on the time scale of nano seconds, but it could be emulated by the progressive decrease of the amorphous volume in the PCM layer as a result of repeated cycling, however, the author has no knowledge about any experimental studies with respect to this issue so that no quantificational conclusions can be drawn at this point. From a purely neuro-scientific computational point of view, the damping of the amplitudes is commonly assumed to have no relevance as the only important characteristics of modeled neuronal behaviour in artificial neural networks is the temporal structure of the neuronal firing activity. Note that both models, the I&F model and the R&F model, which are widely used to model the neuronal behaviour in artificial neural networks for computational studies, do not even consider the generation of an actual spike (see section 4.4). Spikes in these computational studies are mere digital events and have no biological resemblance whatsoever [51, 56]. Thus, the presented concept of an artificial PCM bursting neuron could perfectly serve for computational study purposes if the temporal structure of the emulated bursting behaviour can be sufficiently manipulated. However, recent research suggests that the actual shape of the APs carries information proposing an 'action potential waveform code' which states that four times more information can be transferred by broader somatic APs, reliably produced in response to higher conductance inputs, than by spike times alone [97].

5.2. The Analogy to Hippocampal Pyramidal Bursting Neurons

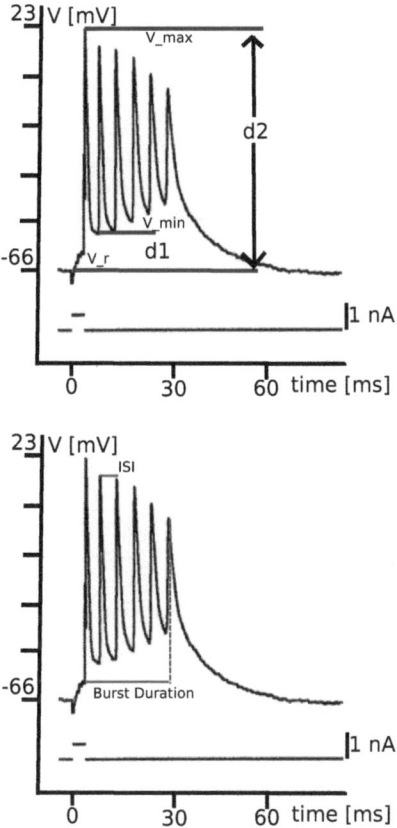

Figure 5.5.: *Parameterizing hippocampal pyramidal neuronal bursting activity.* **Top:** *The relative amplitude can be calculated from the minimum and maximum values of the membrane potential (V_{min}, V_{max}) during the burst. This amplitude is damped which cannot be emulated by a common resistor because its resistance would have to be variable on a nano second time scale.* **Bottom:** *The ratio of the ISIs within a burst to the burst duration is the most important temporal characteristics of a burst and needs to be considered when fitting the voltage oscillations. [modified from [56]]*

Chapter 5: A PCM Bursting Neuron

While this information is preserved during synaptic integration, it is still unclear whether this information can survive axonal conduction and directly influences synaptic transmission. In order to fit the temporal structure of neuronal bursting activity, it is necessary to examine α_{burst}, the ratio of the burst duration to the ISIs within this burst, which is illustrated in the bottom part of figure 5.5 and can be estimated to

$$\alpha_{burst} = \frac{\tau_{burst}}{\tau_{ISI}} \approx \frac{24.1\,\text{ms}}{4.2\,\text{ms}} \approx 5.7$$

A similar estimation for the PCM bursting neuron yields

$$\alpha_{PCM} = \frac{T_{on}}{\tau_{ISI}} \approx \frac{400.00\,\text{ns}}{43.28\,\text{ns}} \approx 9.2$$

(see Figure 5.2). There are two possibilities to fit the temporal structure of the PCM bursting neuron: (1) the time T_{on} over which V_A is applied can be reduced or (2) the time between two spikes can be increased which is determined by the time constant τ_C for capacitance charge and discharge during the transitions from the a-off into the a-on state and vice versa. While varying T_{on} is trivial, τ_{ISI} depends on R_L, R_{DUT}^{on} and C and should only be varied through the variation of C since changing the ratio of R_L to R_{DUT}^{on} would also change the relative amplitude of the voltage oscillations. The needed capacitance C can be calculated from

$$\tau_{ISI} = 5 \cdot \left(R_{DUT}^{on} \cdot C + R_L \cdot C \right)$$

while the needed τ_{ISI} can be estimated to $400\,\text{ns}/5.7 = 70.18\,\text{ns}$, so that

$$70.18\,\text{ns} \stackrel{!}{=} 5 \cdot (3\,\text{k}\Omega \cdot C + 17.34\,\text{k}\Omega \cdot C)$$
$$\Rightarrow C = \frac{70.18\,\text{ns}}{101.7\,\text{k}\Omega} = 0.69\,\text{pF}$$

The factor 5 comes from the fact that the capacitance has charged/discharged 99.3% after the time $5\tau_C$ after which the charging/discharging process can be regarded as complete. Note that it is not necessary to have the capacitance completely charged/discharged but if this isn't the case, e.g. if one choses to define the time between two spikes as

$$\tau_{ISI} = R_{DUT}^{on} \cdot C + R_L \cdot C,$$

the capacitance has charged/discharged 63.2% so that the voltage across the parallel PCM cell has not completely build up which has to be taken into consideration when fitting the relative amplitude of the oscillations.

In general, smaller capacitances are desirable in order to increase the oscillation frequency f_{osc} of the system. When decreasing the capacitance, two factors must be considered: (1) the ratio of T_{on} to τ_{ISI} must be constant, i.e. T_{on} has to be changed accordingly and (2) the influence of the delay time τ_d when τ_{ISI} approaches values of the same order of magnitude as

5.2. The Analogy to Hippocampal Pyramidal Bursting Neurons

τ_d. When τ_{ISI} becomes smaller than τ_d, the transition from the a-off state into the a-on state is governed by τ_d and the PCM remains in the a-off state for the period of τ_d even though V_T is already applied. This would create kind of a plateau instead of a sharp spike so that the PCM neuron loses it's similarity to its biological role model. For values of V_A of about 1.85 V, a delay time of $\tau_d \approx 5$ ns has been demonstrated for GeTe$_6$ [79] which gives a rough estimation of how small the capacitance can be chosen. A transition time τ_{ISI} of e.g. 10 ns, which is still sufficiently larger than $\tau_d = 5$ ns, can be achieved with a capacitance $C \approx 98$ fF. Note that T_{on} has to be scaled accordingly to $T_{on} \approx 57$ ns in oder to preserve the original temporal structure of the artificial burst. Even smaller capacitances might be reached if τ_d could be reduced. It has been demonstrated that τ_d can be decreased if the PCM layer is still electrically excited before a threshold switch occurs [37]. Once the PCM layer is in the a-on state, it relaxes back into the a-off state and another threshold switch occurs after the capacitance has charged and V_{DUT} reaches V_T. If the time constant for the capacitance charge is small enough for the PCM layer to be still partially electrically exited from the threshold switching event before, the delay time for the following switching event is shorter than the delay time for the earlier switching event, thus, even smaller values of τ_{ISI} can be reached without τ_d causing plateaus instead of sharp spikes. In addition to the variation of τ_{ISI} through C, f_{osc} can be increased even further by increasing V_A, whereas T_{on} must be changed accordingly again in order to preserve the original temporal structure of the burst.

In summary, the relative amplitude A_{osc} and the temporal structure of the voltage oscillations can be tuned to fit biological bursting activity. While the A_{osc} depends on the ratio of R_L to R_{DUT}^{on} and can be tuned by choosing the correct value for R_L, the temporal structure depends on the ratio of T_{on} to τ_{ISI}. T_{on} can be trivially changed, whereas τ_{ISI} depends on the time for capacitance charge and discharge and thus, the value of C. Furthermore, the oscillation frequency f_{osc} can be tuned independently from A_{osc} by changing V_A.

Considering the above discussion and the ways in which the physical parameters influence the characteristics of an artificial burst, it is possible to find analogies between the physical parameters of the PCM bursting neuron and physiological parameters influencing the burst of hippocampal pyramidal LTB II and LTB III neurons. These neurons have two parameters that influence the two most important characteristics of a burst, namely the ISIs within a burst and the burst duration: (1) the extracellular Ca^{2+}-concentration $[Ca^{2+}]_o$ and (2) the Ca^{2+}-influx. The extracellular Ca^{2+}-concentration $[Ca^{2+}]_o$ regulates the slow Na$^+$-current $I_{Na,slow}$ which facilitates bursting activity. The exact physiological mechanisms are not yet resolved but it is believed that extracellular Ca^{2+}-ions block slow Na$^+$-currents so that an increase of $[Ca^{2+}]_o$ leads to a decrease of $I_{Na,slow}$ and therefore to a shorter burst [58, 61]. The Ca^{2+}-influx regulates the activation of Ca^{2+}-gated K$^+$-channels whose activation therefore depends on how much Ca^{2+}-ions enter the cell. A lower Ca^{2+}-influx results in less activated K$^+$-channels which antagonize the generation of a spike, so that a lower Ca^{2+}-influx conse-

Chapter 5: A PCM Bursting Neuron

quently results in shorter ISIs and thus, higher burst frequencies [58, 61]. The ISIs within an artificial burst of a PCM bursting neuron are determined by the capacitance C and the applied voltage V_A, whereas the ISIs in hippocampal pyramidal LTB II and LTB III neurons are determined by the Ca^{2+}-influx. A decrease of C as well as an increase of V_A has a similar effect as a decrease of Ca^{2+}-influx and results in shorter ISIs within the bursts so that the capacitance C and the inverse of V_A of the PCM bursting neuron can be directly linked to its correlated physiological parameter, the Ca^{2+}-influx. In the simple model, the dimensionless parameter c determines the modeled systems ability to generate spikes during a burst and therefore directly influences the ISIs within the burst. An increase of c leads to a decrease of the ISIs and thus, c influences the ISIs within a burst in the same way as its physical analogy V_A but inversely to its physiological analogy and its physical analogy C. The burst duration of an artificial burst, i,e the time T_{on} over which V_A is applied, can be directly linked to the extracellular Ca^{2+}-concentration $[Ca^{2+}]_o$, which determines the burst duration of hippocampal pyramidal LTB II and LTB III neurons. An increase in $[Ca^{2+}]_o$ and a decrease of T_{on} results in an increased burst duration. The dimensionless parameter d of the simple model determines how much ability of the system to spike is consumed for each spike and thus, influences the burst duration of the modeled system. An increase of d leads to a decreased burst duration and thus, d influences bursting activity in the same way as its physiological analogy but inversely compared to its physical analogy. Additionally, both characteristics, the ISIs and the burst duration, are represented by the recovery constant a of the simple model which generally determines the over all pace of the modeled system and thus, changes the ISIs as well as the burst duration in a way such that the temporal structure of the modeled system is preserved. Table 5.1 summarizes the important parameters of the physical and biological systems as well as their influence on bursting activity and their correlation with each other as well as with the parameters of the simple model. **Complementary to the discussions above,**

Burst Characteristics	Physiological Parameter	Physical Parameter	Simple Model Parameter
ISI ↓	Ca^{2+}-influx ↓	C ↓ or V_A ↑	c ↑
Burst Duration ↓	$[Ca^{2+}]_o$ ↑	T_{on} ↓	d ↑

Table 5.1.: *Summary of the important parameters influencing the characteristics of bursting activity. Both bursting systems, the biological and the physical one, have specific parameters influencing the burst in the same way, thus, creating a direct analogy to one another. Additionally, these parameters have direct analogies in the simple model.*

the next section presents a simplified LTSpice simulation of the electrical circuit which partly verifies the calculated values and gives rise to a slightly modified method for the emulation of neuronal bursting activity.

5.3. Simulation of a PCM Bursting Neuron

For the purpose of verification of the calculated values for the emulation of hippocampal pyramidal neuronal bursting activity, an LTSpice simulation of the electrical circuit, in which the electrical switching behaviour of the PCM cell is implemented in an extremely simplified fashion, verifies the calculated values for C and R_L that have to be chosen in order to create the desired bursting behaviour. In the first instance, the electrical circuit and the voltage oscillations presented in [78] were reproduced to demonstrate that the simplified implementation of the electrical switching behaviour of the PCM cell can, nevertheless, lead to voltage oscillations as experimentally measured. Figure 5.6 illustrates the electrical circuits simulating the voltage oscillations consisting of the actual circuit in which voltage oscillations are observed (bottom) and a single voltage source for the purpose of implementing the electrical switching behaviour (top). The actual circuit contains a voltage source, a load resistance and

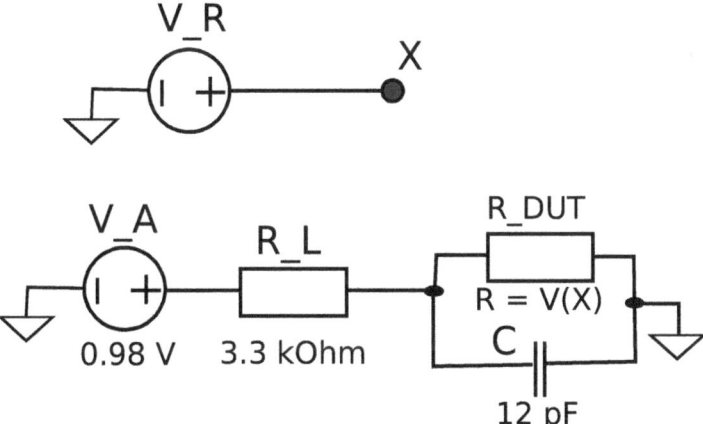

Figure 5.6.: *Schematic illustration of the electrical circuit presented in [78] which is used to induce voltage oscillations. The electrical switching behaviour of the PCM cell is implemented (in highly simplified fashion) by a voltage dependent resistor which switches its resistance at V_T and V_H between 1 kΩ and 3 MΩ for the a-on and a-off state, respectively*

Chapter 5: A PCM Bursting Neuron

the PCM cell in series, whereas the PCM cell is implemented by a voltage dependent resistor and a parallel capacitance. the simulation implements the electrical switching behaviour of PCM cells with the help of an additional voltage source which switches between two voltage values that represent the resistances of the a-off and a-on state. The voltage source switches between 1 kV and 3 MV after a delay time of 100 ns - equivalent to the experiment in [78] as can be seen in Figure 5.2 - with switching times, i.e. rise and fall times, of 0.01 ns and 0.01 ns, respectively, whereas the time for the higher voltage value was chosen to be 32 ns. Ten cycles are generated with a period of 40 ns, so that the simulated burst duration is 400 ns, as in [78]. Figure 5.7 illustrates the simulated voltage oscillations which are in good agreement with the experimental measurements presented in [78] which are illustrated in 5.2. Figure 5.8 illustrates simulated voltage oscillations of the same electrical circuit with the

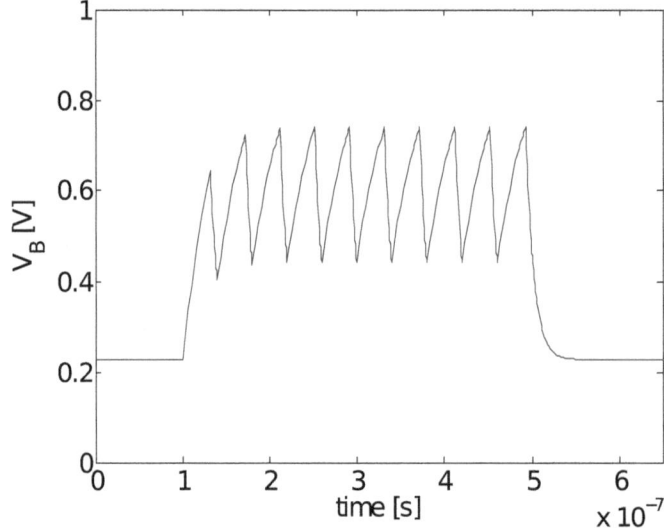

Figure 5.7.: Highly simplified LTSpice simulation of experimentally measured voltage-controlled relaxation oscillations. The trace shows that the experimentally measured voltage oscillations in [78] can be reproduce with the highly simplified implementation of the electrical switching behaviour of the PCM cell.

calculated values of the previous section: $V_A = 1.85\,\text{V}$, $R_L = 17.34\,\text{k}\Omega$, $C = 98\,\text{fF}$, $R_{DUT}^{off} = 80\,\text{M}\Omega$, $R_{DUT}^{on} = 3\,\text{k}\Omega$, $T_{on} = 57\,\text{ns}$ and $\tau_{ISI} = 10\,\text{ns}$. The simulated voltage oscillations show the correct relative amplitude and the correct temporal structure, whereas the shape of the spikes is

5.3. Simulation of a PCM Bursting Neuron

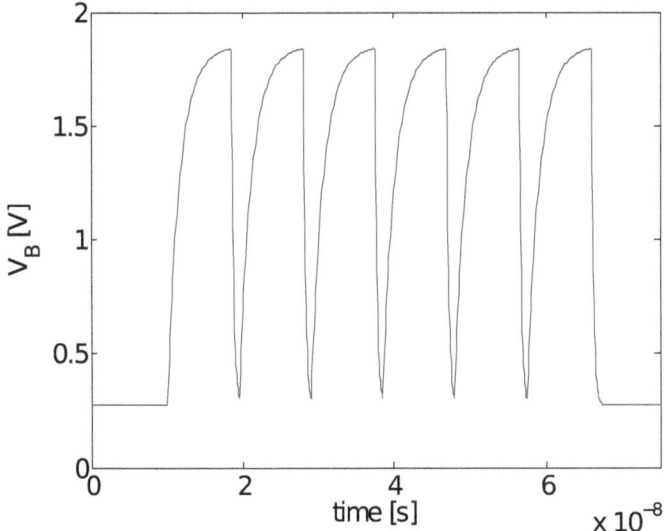

Figure 5.8.: Highly simplified LTSpice simulation of voltage oscillations across a PCM cell for neuronal bursting activity emulating purposes. The relative amplitude and the temporal structure for the calculated values are in good agreement with the biological trace of hippocampal pyramidal neurons in Figure 5.5.

based on the exponential charging and discharging characteristics of the capacitance. Note that no ultimate conclusions can be drawn about the correctness of the temporal structure for the calculated values due to the oversimplified implementation of the electrical switching behaviour of the PCM cell, however, the simulation verifies the correctness of the calculated values for the relative amplitude A_{osc}. In order to appreciate this simplification, the transient effects, that have been mentioned in chapter 3, section 3.2.1, are briefly explained:

One of these transient effects occurs when the a-on state collapses after the voltage across the PCM cell drops below the holding voltage V_h, the minimum voltage which has to be applied to the PCM cell in order to sustain the a-on state. The conductivity of the PCM cell decays over a time τ_r, the transient recovery time, over which the voltage across the PCM cell increases and the a-off state is recovered in a continuous fashion. Another transient effect can be observed when switching off the applied voltage in the a-on state and reapplying a voltage lower than V_{th} and higher than V_h after an interruption time t_s. It has been demonstrated that there is a maximum interruption time $t_{s,max}$ after which the a-on state can not be sustained

Chapter 5: A PCM Bursting Neuron

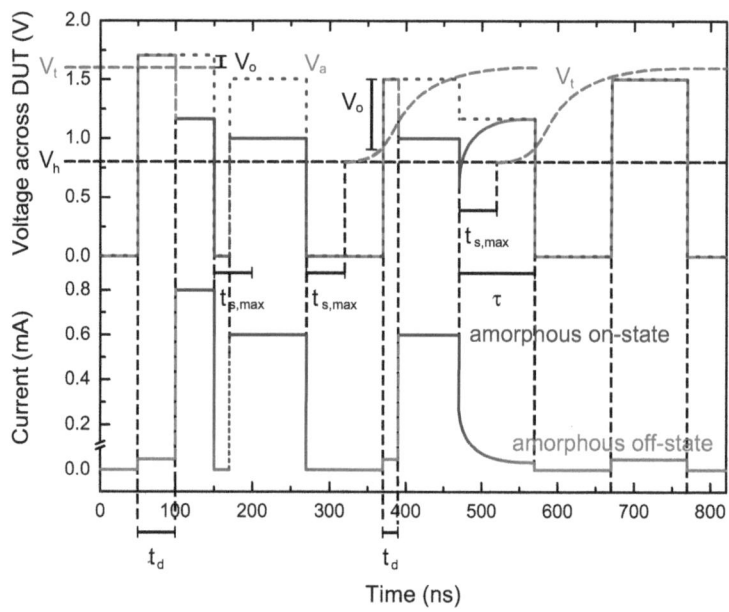

Figure 5.9.: Schematic illustration of the transient effects during electrical switching of PCM. The applied voltage V_A contained a sequence of four pulses (dotted blue line) and was applied to a PCM device in series with a load resistor. Each of the three transient effects discussed in the text can be observed during the pulse sequence. (1) The collapse of the a-on state upon voltage drop below V_h, (2) the switching event into the a-on state upon pulse application for voltages lower than V_{th} but higher than V_h and corresponding to that, the dependence of V_{th} on the the time between two switching events and (3) the dependence of τ_d on V_A, or more precisely on $V_o = V_A - V_{th}$. Detailed explanations are left out at this point for lack of space but can be found in the text. [taken from [36]]

by applying a voltage between V_{th} and V_h so that V_{th} has to be overcome again in order to induce switching into the a-on state. Another transient effect occurs at the threshold switch itself, when V_A is higher than V_{th}. It has been observed that τ_d decreases exponentially with the overvoltage V_o, i.e. the difference between V_A and V_{th}. Figure 5.9 illustrates the transient effects discussed above. The applied voltage V_A contained a sequence of four pulses (dotted

5.3. Simulation of a PCM Bursting Neuron

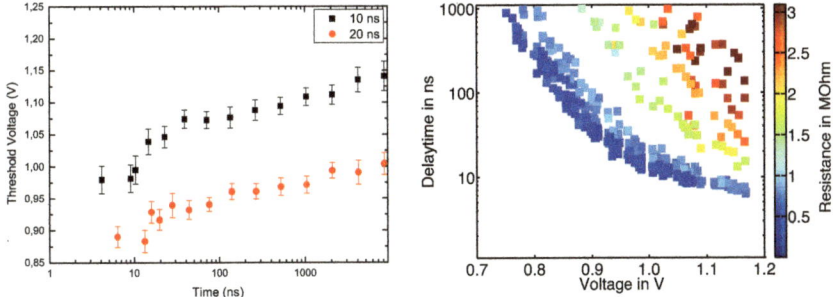

Figure 5.10.: *Experimental measurements illustrating the transient effects on the threshold voltage V_{th} and the delay time τ_d. **Left:** Dependence of V_{th} on the time between two voltage pulses that induce threshold switching. The threshold voltage V_{th} recovers to its original value with increasing time between the two pulses and furthermore, depends on the pulse rise times which has been demonstrated for two exemplarily values of 10 ns and 20 ns. **Right:** Dependence of τ_d on V_A for various initial resistance values. The delay time τ_d is a decreasing function of the applied voltage V_A and thus, V_o, as well as of the initial resistance of the PCM. [taken from [37, 38]]*

blue line) and was applied to a PCM device in series with a load resistor. The first pulse, slightly above V_{th}, induces the transition from the a-off state into the a-on state after a delay time τ_d which is accompanied by a dramatic increase of the current. Due to the dramatic change in conductivity and thus, resistance of the PCM device, the total voltage drop shifts towards the resistor resulting in a voltage snapback across the PCM device upon switching into the a-on state. After an interruption time t_s shorter than $t_{s,max}$, a second voltage pulse between V_{th} and V_h is applied to order to sustain the a-on state, i.e. the PCM enters the a-on state upon pulse application without crossing the original value of V_{th}, which means that V_{th} for the second switching event is reduced compared to the original value for the first switching event. For interruption times longer than $t_{s,max}$, V_{th} starts to recover to its original value. The third pulse, exceeding V_{th} by an overvoltage V_o, results in a decreased delay time τ_d for the switching event. After the transition into the a-on state, the voltage across the PCM device drops below V_h leading to the continuous recovery of the a-off state during the recovery transient time τ_r. The fourth pulse is applied after V_{th} has almost completely recovered to its original value so that the voltage across the PCM device does not exceed V_{th} and the PCM remains in the a-off state. [36] Complementary to the above explanations,

Figure 5.10 illustrates the dependence of τ_d on V_A [38] as well as the dependence of V_{th} on the time between two switching events [37].

The implementation of the electrical switching behaviour ignores the transient effects discussed above and considers only the transition from a high resistance state into a low resistance state and vice versa. The time after which these transitions occur are chosen manually based on the calculation for the charging and discharging time of the capacitance after which the voltage values V_T and V_H are reached which leads to the transitions.

5.4. An Overall View

The assumed importance of neuronal bursting behaviour with respect to the processing of information in the human brain makes it desirable to find possibilities to experimentally emulate this behaviour in order to utilize these emulations within complete artificial neural networks which can be used to experimentally complement simulations and computational studies. Voltage-controlled relaxation oscillations as the result of the inherent electrical switching characteristics of PCMs are proposed to emulate biological bursting activity of hippocampal pyramidal LTB II and LTB III neurons. In order to utilize the oscillation behaviour as an artificial burst, the physical parameters of the artificial PCM bursting neuron are analyzed in terms of how they influence artificial bursting activity and can be directly linked to physiological parameters which influence biological bursting activity in the same way, as well as to parameters of the simple model which is used to model neuronal bursting activity in computational studies. A highly simplified LTSpice simulation verified the correctness of the calculate values for the physical parameters for the emulation of the relative amplitude A_{osc}. **The next chapter concludes this thesis with a brief summary of the documented work and gives an outlook on future ideas, some of them being complementary to, others being independent on the concept of this thesis.**

CHAPTER 6

An Outlook on the Future

The human brain represents a remarkable information processing architecture providing several unique features which emphasize this architecture's superiority with respect to certain computational tasks, especially those involving any kind of pattern recognition, in comparison with today's computational standards based on binary logic. The desire to keep up the trend of computational progress, described by Moore's law, gives reason to research possibilities to emulate the brain's superior functionalities in order to overcome the problems caused by the increasing complexity of today's computational tasks as well as the barriers posed by the nano era. These possibilities are commonly based on nano electronic devices used in artificial neural networks emulating biological functionalities which are thought to be the reason for the exceptionally high performance of the brain considering certain computational tasks.

While most of the current research is based on the emulation of synaptic plasticity (see chapter 3), effects which are commonly thought to be the underlying ones for emergent phenomena such as memory and learning, the concept of this thesis follows the idea to emulate neuronal bursting activity, a specific firing pattern exhibited by some neurons in the brain, which is thought to play important roles for the communication between neurons for various reason (see chapter 4), thus, making its emulation feasible. The emulation of neuronal bursting activity is proposed to be adopted by an electrical circuit, containing a common resistor connected in series to a PCM device with a parallel capacitance, which exhibits voltage oscillations that can be manipulated to match neuronal bursting activity of hippocampal pyramidal neurons (see chapter 5). This electrical circuit could possibly serve as an artificial neuronal bursting device within complete artificial neural networks following the interest to complement biological or computational studies, which consider the importance of neuronal bursting activity, by experiments.

Chapter 6: An Outlook on the Future

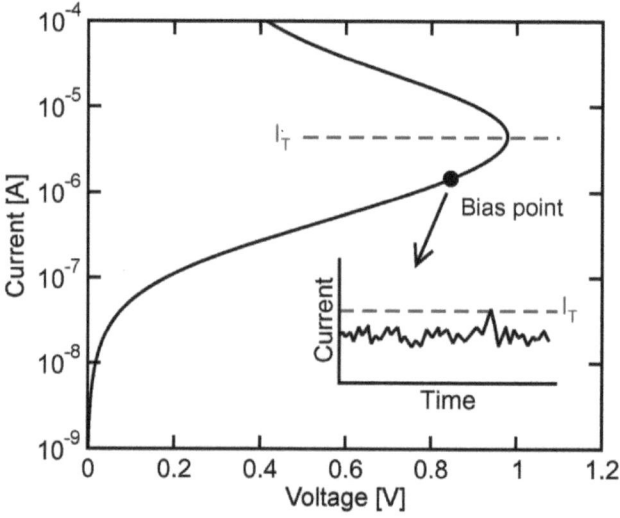

Figure 6.1.: Physical picture of threshold switching induced by current fluctuations in the subthreshold regime. When a PCM cell is biased slightly below the threshold voltage, resulting current fluctuations due to the presence of $1/f$ noise induce threshold switching upon crossing the value of I_T. [taken from [88]]

Complementary to the documented concept, additional work steps could be carried out in order to provide experimental verification for the presented conceptual work. The Phase-Change Electrical Tester (PET) - the gadget with which the I. Physikalische Institut (IA) of the RWTH Aachen University contacts its PCM cells and carries out electrical switching experiments - could be modified, i.e. a load resistance could be built in which would enable the induction of voltage-controlled relaxation oscillations. Furthermore, experiments could be carried out which follow the interest to find out how exactly the parasitic capacitance depends on the different factors such as the geometry of the cell or the contact needles, in order to determine the most convenient way to manually control the capacitance with the least necessary effort. Additionally, experiments quantifying the progressive collapse of the threshold voltage, which could be used to emulate damped bursting behaviour, could be carried out in order to provide additional analogy between artificial and biological bursting activity.

Independent on the documented work, the author beliefs that it is potentially interesting

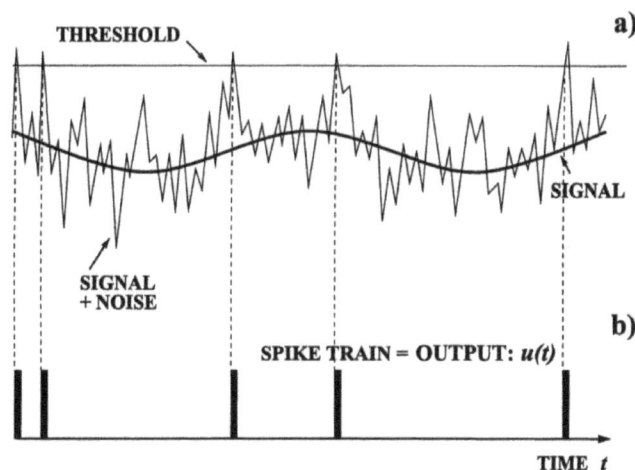

Figure 6.2.: *The effect of stochastic resonance is rooted in three minimal ingredients: (1) a source of background noise, (2) a generically weak coherent input and (3) a characteristic sensory barrier or threshold that the system typically has to overcome in order to perform its useful task. [taken from [89]]*

to investigate the possibilities to emulate a biological mechanism called stochastic resonance through the threshold switching mechanism in PCM. The reason for this is founded in the striking similarity of the physical pictures of both mechanisms given by Lavizzari et al. [88] and Haenggi [89], as illustrated in Figure 6.1 and Figure 6.2, respectively. The physical picture of threshold switching given by Lavizzari et al. explains threshold switching as the result of current fluctuations in the subthreshold regime causing current spikes that exceed the threshold current value I_T triggering threshold switching which suddenly decreases the cell's resistance to the a-on state value. The physical picture of stochastic resonance, an effect which can be found in different biological systems [90–96], is based on three minimal ingredients: (1) a source of background noise, (2) a generically weak coherent input and (3) a characteristic sensory barrier or threshold that the system has to overcome to perform its useful task. A comparison of both physical pictures leads to the notion that the effect of stochastic resonance, which plays an important role in improving biological information processing [89], could be emulated by the threshold switching mechanism in PCM cells. In a PCM device, the source of background noise is provided by biasing the device slightly below the threshold voltage which leads to current fluctuations in the subthreshold regime.

Chapter 6: An Outlook on the Future

The sensory barrier, which has to be overcome, is represented by the threshold of the PCM and the useful task of the system, which is performed upon crossing that sensory barrier, is represented by the dramatic increase in conductivity of the PCM cell upon crossing the threshold voltage and switching from the a-off state into the a-on state. The weak coherent signal, which is supposed to be detected, is provided by the environment in which the device is supposed to operate, and would have to be measured and converted into some sort of voltage signal which is applied to the device in addition to the constant subthreshold bias. Hence, a PCM device might be utilized to detect some kind of weak signals based on the fact that current fluctuations are able to trigger threshold switching in the device, thus, successfully emulating the effect of stochastic resonance.

Overall concluding, the material presented in this thesis suggests that, with respect to the intention to understand how the human brain works, phase-change materials are a suitable candidate for experimental emulations of different biological functionalities which are commonly thought to be the underlying mechanisms for the emergent phenomena which make the human brain that highly remarkable and equally puzzling thing that it is.

APPENDIX A

Quantification of the Membrane Potential

The Nernst Equation

The equilibrium potential E_X for any ion X, i.e. the membrane potential at which there is no net flux of the ion species across the cell membrane, can be calculated from an equation derived in 1888 from basic thermodynamical principles by the German physical chemist Walter Nernst:

$$E_X = \frac{RT}{zF} \cdot ln\frac{[X]_o}{[X]_i},$$

where R is the gas constant, T the temperature (in degrees Kelvin), z the valence of the ion, F the Faraday constant and $[X]_o$ and $[X]_i$ are the ion concentrations outside and inside the cell respectively. With $RT/F = 25$ mV at 25°C (room temperature) and a converting factor of 2.3 from natural logarithms to base 10 logarithms, the Nernst equation can also be written as:

$$E_X = \frac{58\,mV}{z} \cdot log\frac{[X]_o}{[X]_i}.$$

Table A.1 shows the distribution of the major ions across a neuronal membrane at rest and the calculated equilibrium potentials of the giant axon of the squid which was examined by Hodgkin and Katz [85]. [13]

Appendix A: Quantification of the Membrane Potential

Species of ion	Concentration in cytoplasm [mM]	Concentration in extracellular fluid [mM]	Equilibrium potential [mV]
K$^+$	400	20	-75
Na$^+$	50	440	+55
Cl$^-$	52	560	-60
A$^-$ (organic anions)	385	-	-

Table A.1.: Distribution of the major ions across a neuronal membrane at rest: the giant axon of the squid [85]. [taken from [13]]

The Goldman Equation

The contributions of different ions to the resting membrane potential can be quantified by the Goldman equation. Although ion fluxes of Na$^+$ and K$^+$ determine the resting potential value, V_m is not equal to E_{Na} or E_K but lies between them. When the membrane potential is determined by two or more species of ions, the influence of each species on the membrane potential is determined not only by its intracellular and extracellular concentrations but also by the permeability P of the membrane to the specific ion species measured in units of velocity, cm/s. The dependence of the membrane potential on ionic permeability and concentration is given quantitatively by the Goldman equation, which applies only when V_m is not changing:

$$V_m = \frac{RT}{zF} \cdot \ln\left(\frac{P_K[K^+]_o + P_{Na}[Na^+]_o + P_{Cl}[Cl^-]_i}{P_K[K^+]_i + P_{Na}[Na^+]_i + P_{Cl}[Cl^-]_o}\right).$$

Qualitatively this means that the greater the concentration of a particular ion species and the greater its membrane permeability, the greater its role in determining the membrane potential. In the limit of one single ionic permeability being much greater than all the others, the Goldman equation reduces to the Nernst equation for that specific ion. For instance, if $P_K \gg P_{Na}$ and P_{Cl}, as is the case in glial cells, the equation becomes:

$$V_m \cong \frac{RT}{F} \cdot \ln\frac{[K^+]_o}{[K^+]_i}.$$

[13]

APPENDIX B

Vocabulary

- **action potential:** An action potential (AP) generated by a neuron is an all-or-nothing event in which the neuron's membrane potential rapidly rises and falls due to the opening and closing of several ion channels integrated in the cell membrane that facilitate the flow of ionic inward and outward currents.

- **active zone:** The active zone is a specific zone of the presynaptic terminal membrane which facilitates the release of neurotransmitter by providing the fusion of vesicles.

- **after potential:** The after potential is a short period of hyperpolarization immediately following the action potential during which the membrane potential drops below the resting potential value.

- **all-or-nothing:** An all-or-nothing event is an event that either happens completely or it doesn't happen at all. The event itself doesn't depend on the strength of the signal which causes this event.

- **amplification:** Amplification is the ability of chemical synapses to evoke large PSPs caused by relatively weak presynaptic currents and is based on the release of a huge amount of neurotransmitter molecules in response to the weak presynaptic currents, thereby facilitating the activation of whole populations of postsynaptic ion channels.

- **axoaxonic synapse:** An axoaxonic synapses connects a presynaptic terminal to a postsynaptic axon and is often inhibitory. (↪ synapse)

- **axodendritic synapse:** An axodendritic synapses connects a presynaptic terminal to a postsynaptic dendrite or its spine and is often excitatory. (↪ synapse)

Appendix B: Vocabulary

- **axon:** The axon, also called nerve fiber, is a long, slender extension of the soma that originates at the axon hillock and functions as the main conducting unit that carries signals away from the soma to other neurons.

- **axosomatic synapse:** An axosomatic synapses connects a presynaptic terminal to a postsynaptic soma and is often excitatory. (↪ synapse)

- **axon hillock:** The axon hillock is a specialized region of the soma that connects to the axon. As the last site in the soma where membrane potentials propagated from synaptic inputs are being transmitted to the axon it functions as an integrator of subthreshold inputs.

- **burst:** ↪ bursting

- **bursting:** Bursting is a dynamic state in which a neuron repeatedly fires discrete groups of consecutive AP's, also referred to as bursts.

- **capacitive membrane current:** The capacitive membrane current is carried by ions that change the net charge stored on a cell membrane.

- **cell body:** The cell body is the metabolic center of the neuron and contains the cell nucleus, i.e. It contains most of the cell's genetic material organized as multiple long linear DNA molecules in complex with a large variety of proteins to form chromosomes.

- **cell membrane:** The cell membrane is a thin polar membrane, the lipid bilayer sheet, which separates the interior of a neuron from the outside environment preventing the diffusion of ions, proteins and other molecules across the cell membrane and is selectively permeable to ions and organic molecules controlling the movement of substances in and out of the neuron.

- **chemical synapse:** ↪ synapse

- **cytoplasm:** The cytoplasm is a gel-like substance within the cell membrane holding all the cell's internal substructures, called organelles, except for the nucleus.

- **dendrite:** The dendrites are the branched extensions of a neuron's soma functioning as the main input receiving apparatus of the neuron, thus, they conduct electrochemical signals received from other neurons to the soma.

- **dendritic tree:** The dendritic tree is the entity of all dendritic extensions and is named after its morphology which manifests itself in extensive tree-like branching.

- **depolarization:** Depolarization is the reduction of charge separation across a neuron's cell membrane leading to a less negative membrane potential.

- **electrical synapse:** ↪ synapse

- **endocytosis:** In general, endocytosis is the process by which cells absorb molecules by engulfing them. In this way, e.g. large polar molecules that cannot pass through the hydrophobic cell membrane can be assimilated.

- **EPSC:** An excitatory postsynaptic current (EPSC) depolarizes a postsynaptic cell membrane in response to excitatory synaptic input.

- **EPSP:** An excitatory postsynaptic potential (EPSP) is the change in the postsynaptic membrane potential caused by depolarization in response to an EPSC.

- **equilibrium potential:** The equilibrium potential for a specific ion species is the membrane potential at which there is no net flux of this ion species across the cell membrane.

- **exocytosis:** In general, exocytosis is the process by which cells direct the contents of secretory vesicles out of the cell membrane and into the extracellular space.

- **gap-junction channel:** A gap-junction channel is the connecting channel between two neurons of an electrical synapse which transmits signals by conducting ionic currents from the presynaptic cell to the postsynaptic cell.

- **glial cell:** Glial cells are non-neuronal cells maintaining homeostasis, form myelin, and provide support and protection for neurons in the brain and in other parts of the nervous system.

- **Hebbian learning:** Hebbian learning is a method of associative learning of the nervous system in which synaptic connections are strengthened as a result of repeated and persistent stimulation of the postsynaptic neuron by a presynaptic neuron.

- **Hebbian theory:** The Hebbian theory, introduced in 1949 by the psychologist Donald O. Hebb, explains the adaption of neurons in the brain during the process of learning. (↪ Hebbian learning)

- **hyperpolarization:** Hyperpolarization is the increase of charge separation across a neuron's cell membrane leading to a more negative membrane potential.

- **interburst frequency:** The interburst frequency is the frequency with which a neuron generates consecutive bursts that are separated by a period of quiescence.

- **interburst interval:** The interburst interval (IBI) is the temporal interval between two consecutive bursts.

Appendix B: Vocabulary

- **interspike interval:** The interspike interval (ISI) is the temporal interval between two consecutive spikes within a burst.

- **intraburst frequency:** The intraburst frequency is the frequency with which a neuron generates consecutive spikes within a single burst.

- **in vitro:** In vitro (latin for "within the glass") is experimentation using components of an organism that have been isolated from their usual biological surroundings in order to permit a more detailed or more convenient analysis than can be done with in vivo experiments. (↪ in vivo)

- **in vivo:** In vivo (latin for "within the living") is experimentation using a whole, living organism as opposed to an in vitro controlled environment. (↪ in vitro)

- **ion channel:** Ion channels are usually integrated in the cell membrane and conduct ions across the cell membrane between the intracellular and extracellular space at extremely rapid rates providing selective permeability of the cell membrane.

- **ion pump:** Ion pumps are usually integrated in the cell membrane and prevent the dissipation of ionic gradients by moving ions *against* their net electrochemical gradient.

- **ionic membrane current:** The ionic membrane current is carried by ions passing an ion channel changing the intracellular and extracellular ion concentrations.

- **IPSC:** An inhibitory postsynaptic current (IPSC) hyperpolarizes a postsynaptic cell membrane in response to inhibitory synaptic input.

- **IPSP:** An inhibitory postsynaptic potential (IPSP) is the change in the postsynaptic membrane potential caused by hyperpolarization in response to an IPSC.

- **leakage conductance:** The leakage conductance is the conductance of the population of all resting channels.

- **leakage current:** The leakage current is carried by ions passing through resting channels that maintain the resting potential.

- **local-circuit current:** The local-circuit current flows through the axon as a result from the potential difference between active and inactive regions of the membrane.

- **long-term depression:** Long-term depression (LTD) is a form of LTP which results in a long lasting decrease of synaptic weight.

- **long-term plasticity:** Synaptic plasticity is categorized as long-term plasticity (LTP - not to be confused with long-term potentiation) when the plastic changes last several minutes or longer and is the result of an increased Ca^{2+}-concentration in the postsynaptic cell.

- **long-term potentiation:** Long-term potentiation (LTP - not to be confused with long-term plasticity) is a form of LTP which results in a long lasting increase in synaptic weight.

- **membrane length constant:** When considering the propagation of action potentials through the axon, the membrane length constant defines the distance after which the change in membrane potential has decayed to $1/e$, i.e. 37% of its initial value.

- **membrane potential:** The membrane potential is the difference in electrical potential between the intracellular and extracellular space of a neuron caused by a separation of charges across the cell membrane.

- **metabolism:** Metabolism is the set of chemical reactions happening in the neuron to sustain life. These processes allow organisms to grow and reproduce, maintain their structures, and respond to their environment.

- **myelin:** Myelin is an electrically insulating material outgrown from a type of glial cell and is usually found wrapped around the axon of a neuron. Those myelin sheaths are essential for the proper functioning of the nervous system, especially for the rapid propagation of action potential through nerve fibers.

- **myelin sheath:** ↪ myelin

- **nerve cell:** A nerve cell is an electrically excitable cell in the nervous system that processes and transmits information by electrical and chemical signaling.

- **neural network:** A neural network is a network or circuit of neurons connected to each other by synapses.

- **neuron:** ↪ nerve cell

- **neuronal integration:** Neuronal integration is the integrative action of a neuron to make the decision whether an action potential is generated or not in response to all synaptic input received.

- **neurotransmitter:** A neurotransmitter is a chemical substance that binds to specific receptors in the postsynaptic cell membrane.

Appendix B: Vocabulary

- **node of Ranvier:** Nodes of Ranvier are the gaps between two consecutive myelin sheaths where the insulation of the axon is interrupted and the cell membrane lays bare.

- **presynaptic terminal:** A presynaptic terminal is a specialized part of a synapse which originates at the branched endings of the axon and connects to other neurons to form a communication site.

- **PSAP:** A presynaptic action potential PSAP is a neuron's AP arriving at the presynaptic terminal.

- **PSC:** A postsynaptic current (PSC) is evoked by synaptic transmission which, in case of chemical synapses, opens postsynaptic ion channels which leads to a postsynaptic influx of ions.

- **PSP:** A postsynaptic potential (PSP) is the change in the postsynaptic membrane potential caused by depolarization or hyperpolarization of the postsynaptic membrane potential in response to a PSC.

- **pulsed pair facilitation:** Pulsed pair facilitation (PPF) is a form of short-term plasticity in which PSPs evoked by a PSAP are increased if the action potential closely follows a prior PSAP.

- **refractory period:** The refractory period is a period immediately following an action potential during which it is not at all possible or just exceedingly difficult to excite a neuron.

- **repolarization:** Repolarization is the increase in charge separation across a neuron's cell membrane which occurs immediately after depolarization driving the membrane potential back to its resting potential value.

- **resting potential:** The resting potential is a neuron's membrane potential value at rest and is determined by the steady state activities of resting ion channels and ion pumps integrated in the cell membrane.

- **saltatory conduction:** In myelinated axons, saltatory conduction is the propagation of action potentials that jump from one node of Ranvier, where the action potential is refreshed, to the next, rapidly increasing the conduction velocity of action potentials.

- **short-term augmentation:** Short-term augmentation (STA) is a form of short-term synaptic enhancement which acts on a time scale of the order of seconds.

- **short-term depression:** Short-term depression (STD) is a form of short-term synaptic inhibition which leads to a decrease in sensibility of the postsynaptic neuron to presynaptic AP's.
- **short-term facilitation:** Short-term facilitation (STF) is a form of short-term synaptic enhancement which acts on the time scale of the order of tens of milliseconds.
- **short-term plasticity:** Synaptic plasticity is categorized as short-term plasticity (STP - not to be confused with short-term potentiation) when the plastic changes last a few seconds or less and is the result of an increased Ca^{2+}-concentration in the presynaptic terminal resulting in an increased probability to release neurotransmitter.
- **short-term potentiation:** Short-term potentiation (STP - not to be confused with short-term plasticity) is a form of synaptic enhancement which lasts tens of seconds to minutes.
- **soma:** ↪ cell body
- **spatial summation:** Spatial summation is a neuron's integrative process of summing up synaptic inputs at different communication sites.
- **spike:** In the context of neural computation, APs are called spikes to emphasize the all-or-nothing character of APs and their treatment as single unitary events which encode information.
- **spike-timing-dependent plasticity:** Spike-timing-dependent plasticity (STDP) is a specific form of LTP, whereas the relative timing between presynaptic and postsynaptic spiking determines whether the synaptic weight is increased or decreased.
- **steady state:** A steady state of a physiological system is a state in which a particular variable is not changing but energy must be continuously added to maintain this variable constant.
- **synapse:** A synapse is the connection between two communicating neurons and can be one of two types, chemical or electrical. A chemical synapse consists of a presynaptic terminal and a postsynaptic cell separated by a synaptic cleft. An electrical synapse connects pre- and postsynaptic neuron via gap-channel junctions.
- **synaptic cleft:** The synaptic cleft physically separates the presynaptic terminal and the postsynaptic cell membrane and is about 20-40 nm wide. It is filled with cellular fluid which serves as a medium to enable presynaptically released neurotransmitter molecules to diffuse across the cleft.

Appendix B: Vocabulary

- **synaptic plasticity:** Synaptic plasticity is the remarkable ability of synapses to undergo functional and structural changes as a result of their history in a neural network, strengthening or weakening their connection strength.

- **synaptic vesicle:** Synaptic vesicles are vesicles present in presynaptic terminals and store various neurotransmitters which can be released into the synaptic cleft by the process of exocytosis. (↪ vesicle)

- **synaptic weight:** The synaptic weight of a synapse is their relative connection strength compared or other synapses in a neural network.

- **temporal summation:** Temporal summation is a neuron's integrative process of summing up consecutive synaptic potentials at the same communication site.

- **threshold potential:** The threshold potential is a neuron's membrane potential value at which an action potential is generated if this value is crossed.

- **vesicle:** A vesicle is a small bubble enclosed by a lipid bilayer which functions as a barrier to prevent the diffusion of ions, proteins and other molecules across the bubble shell.

List of Figures

2.1. A typical neuron's morphology. 5
2.2. Several types of stimuli control the opening and closing of ion channels. 7
2.3. The ionic permeability of the cell membrane is provided by integrated ion channels. 8
2.4. The membrane potential results from a charge separation across the cell membrane. 9
2.5. The change of the membrane potential during the generation of an AP can be divided into five phases. 11
2.6. Equivalent electrical circuit representing a neuronal extension, e.g. a neuron's axon. 13
2.7. The change in membrane potential in a passive neuronal extension decays with distance. 14
2.8. APs in myelinated fibers are periodically refreshed at the nodes of Ranvier. 15
2.9. Chemical synaptic transmission involves several steps. 19
2.10. The cycling of synaptic vesicles at nerve terminals involves several distinct steps. 20
2.11. Synaptic contact can occur at the soma, the dendrites or the axon of postsynaptic neurons. 22
2.12. The two most common morphologic types of synapses in the brain are Gray type I and type II. 23
2.13. Simulation of an experiment that shows paired pulse facilitation (PPF) which occurs at many synapses. 26
2.14. The direction of synaptic weight change depends on the postsynaptic membrane potential. 27
2.15. The synaptic weight changes depending on the relative spike timing between presynaptic and postsynaptic neuron. 28

3.1. Operation principle of the artificial neural network chip. 34

List of Figures

3.2. Measurement of the postsynaptic membrane potential showing STD and STF and of the STDP modification function for a single arbitrary synapse. 35
3.3. Programming of a PCM cell is done with voltage pulses. 38
3.4. I-V characteristics of set and reset state. 39
3.5. Interconnection scheme of PCM synapses in a 3D stackable cross-point structure. 40
3.6. The continuous change in synaptic weight can be emulated by a fine control of the PCM cell's resistance. 41
3.7. Implementation of STDP through a specifically engineered pulse sequence. . . 43
3.8. Schematic Illustration of an artificial neuron with three basic elements. 44

4.1. ELL pyramidal neurons show differential responses to local (prey-like) and global (communication-like) stimuli. 48
4.2. ELL neural network simulations involving global inhibitory feedback show differential responses to local and global stimuli. 49
4.3. Bursting activity can result from the somatic-dendritic interplay. 50
4.4. Hippocampal pyramidal neurons can be categorized into five different classes. . 52
4.5. Bursting can be with increasing or decreasing interspike intervals (ISIs) 53
4.6. Bursts can be used for selective communication via resonance. 56
4.7. A neuron's subthreshold membrane potential oscillation lead to resonance and frequency preference. 58
4.8. Neuronal activity modeled with the leaky Integrate-and-Fire model. 60
4.9. Neuronal activity modeled with the Resonate-and-Fire model. 62
4.10. Neuronal activity modeled with the simple model. 63

5.1. Measured I-V characteristics for a the PCM cell in the amorphous state. 66
5.2. Measured voltage oscillation across the PCM cell as a function of time. 67
5.3. Measured oscillation frequency for different sleep times as a function of the bias voltage. 68
5.4. Schematic illustration of the oscillation mechanism in the PCM cell. 69
5.5. Parameterizing hippocampal pyramidal neuronal bursting activity. 73
5.6. Schematic illustration of the electrical circuit presented in [78] which is used to induce voltage oscillations. 77
5.7. Highly simplified LTSpice simulation of experimentally measured voltage-controlled relaxation oscillations. 78
5.8. Highly simplified LTSpice simulation of voltage oscillations across a PCM cell for neuronal bursting activity emulating purposes. 79
5.9. Schematic illustration of the transient effects during electrical switching of PCM. 80

List of Figures

5.10. Experimental measurements illustrating the transient effects of the threshold voltage and the delay time. 81

6.1. Physical picture of threshold switching induced by current fluctuations. 84
6.2. Physical picture of stochastic resonance. 85

List of Figures

List of Tables

2.1. Properties of electrical and chemical synapses. 17
2.2. Summary of the morphological differences between Gray type I and type II synapses. 24
5.1. Summary of the important parameters influencing the characteristics of bursting activity. 76
A.1. Distribution of the major ions across a neuronal membrane at rest: the giant axon of the squid. 88

List of Tables

Bibliography

[1] G.E. Moore (1965). "Cramming more components onto integrated circuits" - Proceedings of the IEEE 1998, Volume: 86, Issue: 1, Page(s): 82 - 85 1

[2] Special Issue - Proceedings of the IEEE 2008, Volume 96, Issue: 2 1

[3] DARPA/DSO BAA08-28, Systems of Neuromorphic Adaptive Plastic Scalable Electronics (SyNAPSE), www.darpa.mil/dso/solicita-tions/BAA08-28.pdf. 1

[4] IBM unveils a new brain simulator - IEEE Spectrum, Nov 2009. 1

[5] D. Kuzum, R.G.D. Jeyasingh, B. Lee, H.S.P. Wong (2011) "Nanoelectronic Programmable Synapses Based on Phase Change Materials for Brain-Inspired Computing" - Nano Letters 2012, Volume 12, Issue 5, Page(s) 2179 - 2186 1, 2, 29, 32, 37, 40, 41, 42, 43

[6] S. Yu, Y. Wu, R.G.D. Jeyasingh, D. Kuzum, H.S.P. Wong (2011) "An Electronic Synapse Device Based on Metal Oxide Resistive Switching Memory for Neuromorphic Computatio" - IEEE transactions on electronic devices 2011, Volume: 58, Issue: 8, Page(s): 2729 - 2737 2, 32

[7] A.K. Friesz, A.C. Parker, C. Zhou, K. Ryu, J.M. Sanders, H.S.P. Wong, J. Deng (2007) "A Biomimetic Carbon Nanotube Synapse Circuit" - Biomedical Engineering Society (BMES) Annual Fall Meeting 2

[8] (2011) T. Ohno, T. Hasegawa, T. Tsuruoka, K. Terabe, J.K. Gimzewski, M. Aono (2011) "Short-term plasticity and long-term potentiation mimicked in single inorganic synapses" - Nature Materials Volume: 10, Page(s): 591 - 595 2, 32

[9] P. Krzysteczko, J. Muenchenberger, M. Schaefers, G. Reiss, A. Thomas (2012) "The Memristive Magnetic Tunnel Junction as a Nanoscopic Synapse-Neuron System" - Advanced Materials Volume: 24, Issue: 6, Page(s): 762 - 766 2, 31

Bibliography

[10] C. Zamarreno-Ramos, L.A. Camunas-Mesa, J.A. Perez-Carrasco, T. Masquelier, T. Serrano-Gotarredona, B. Linares-Barranco (2011) "On spike-timing-dependent-plasticity, memristive devices, and building a self-learning visual cortex" - Frontiers in Neuroscience 2011, Volume: 5, Issue: 26 2, 31

[11] S.H. Jo, T. Chang, I. Ebong, B.B. Bhadviya, P. Mazumder, W. Lu (2010) "Nanoscale Memristor Device as Synapse in Neuromorphic Systems" - Nano Letters 2010, Volume: 10, Issue: 4, Page(s): 1297 - 1301 2, 31

[12] A. Afifi, A. Ayatollahi, F. Raissi (2009) "Implementation of biologically plausible spiking neural network models on the memristor crossbar-based CMOS/nano circuits" - ECCTD 2009, Page(s): 563 - 566 2, 31

[13] E.R. Kandel, J.H. Schwartz, T.M. Jessel, (2000) "Principles of Neural Science - Fourth Edition" 3, 4, 5, 6, 7, 8, 9, 10, 12, 13, 14, 15, 16, 17, 18, 19, 20, 21, 22, 23, 24, 26, 50, 87, 88

[14] Internet article about the "Aktionspotential" in the german Wikipedia, "http://de.wikipedia.org/wiki/Aktionspotential" (reviewd at 23/07/2012). 11

[15] N. Doidge (2007) "The brain that changes itself: Stories of personal triumph from the frontiers of brain science" - United States Viking Press 2007, Page(s): 427 24

[16] D.O. Hebb (1949) "The organization of behavior" New York: Wiley and Sons. 24

[17] D. Purves, G.J. Augustine, D. Fitzpatrick, W.C. Hall A.S. LaMantia, L.E. While (2012) "Neuroscience - Fifth Edition - Sunderland, MA: Sinauer Associates, Inc. 25

[19] C.S. Stephens, J.F. Wesseling (1999) "Augmentation Is a Potentiation of the Exocytotic Process" - Neuron, Volume: 22, Issue: 1, Page(s): 139 - 146 25

[18] R.S. Zucker, W.G. Regehr (2002) "Short-Term Synaptic Plasticity". - Annual Review of Physiology 2002, Volume: 64, Page(s): 355 - 405 25, 26

[20] X. Jianhua, H. Liming, W. Ling-Gang (2007) "Role of Ca^{2+}-channels in short-term synaptic plasticity" - Current Opinion in Neurobiology, Volume: 17, Issue: 3, Page(s): 352 - 359 25

[21] S. Ben Achour, O. Pascaul (2010) "Glia: The many ways to modulate synaptic plasticity" - Neurochemistry International Volume: 57, Issue: 4, Page(s): 440 - 445 25

[22] A. Artola, S. Broecher, W. Singer (1990) "Different voltage-dependent thresholds for inducing long-term depression and long-term potentiation in slices of rat visual cortex" - Nature, Volume: 347, Page(s): 69 - 72 25

Bibliography

[23] A. Artola, W. Singer (1993) "Long-term depression of excitatory synaptic transmission and its relationship to long-term potentiation" - Trends in Neuroscience 1993, Volume: 16, Issue: 11, Page(s): 480 - 487 27

[24] J. Lisman, (1989) "A mechanism for the Hebb and the anti-Hebb processes underlying learning and memory" - Proceedings of the National Academy of Sciences of the USA, Volume: 86, Issue: 23, Page(s): 9574 - 9578 25

[25] G.Q. Bi, M.M. Poo (1998) "Synaptic Modifications in Cultured Hippocampal Neurons: Dependence on Spike Timing, Synaptic Strength, and Postsynaptic Cell Type" - The Journal of Neuroscience 1998, Volume: 18, Issue: 24, Page(s): 10464 - 10472 27, 28, 41, 42, 43

[26] Medline Plus Medical Encyclopedia 27

[27] J. Schemmel, D. Bruederle, K. Meier, B. Ostendorf (2007) "Modeling Synaptic Plasticity within Networks of Highly Accelerated I&F Neurons" - IEEE ISCAS 2007, Page(s): 3367 - 3370 32, 34, 35

[28] J. Schemmel, A. Gruebl, K. Meier, E. Mueller (2006) "Implementing Synaptic Plasticity in a VLSI Spiking Neural Network Model" - IJCNN 2006, Page(s): 1 - 6 35

[29] M. Wuttig (2005) "Phase-Change Materials: Towards a universal memory?" - Nature Materials 2005, Volume: 4, Page(s): 265 - 266 37

[30] Sousa, V. ; Perniola, L. ; Vuillaume, D. ; DeSalvo, B. (2011) "Phase change memory for synaptic plasticity application in neuromorphic systems" - Neural Networks (IJCNN), The 2011 International Joint Conference on July 31 2011-Aug. 5 - 619 - 624 37

[31] C.D. Wright, Y. Liu, K.I. Kohary, M.M. Aziz, R.J. Hicken (2011) "Arithmetic and Biologically-Inspired Computing using Phase-Change Materials" - Advanced Materials 2011, Volume: 23, Issue: 30, Page(s): 3408 - 3413 37

[32] C. Koch (1997) "Computation and the Single Neuron" - Nature 1997, Volume: 385, Page(s): 207 - 210 37

[33] G. Snider (2008) "Cortical computing with memristive nanodevices" - SciDAC Review, 2008 37

[34] H.S.P. Wong et al. (2010) "Phase Change Memory" - Proceedings of the IEEE 2010, Volume 98, Page(s): 2201 - 2227 37, 38, 39

[35] G.W. Burr et al. (2010) "Phase change memory technology" - Journal of Vacuum Science and Technology B: Microelectronics and Nanometer Structures 2010, Volume: 28, Issue: 2, Page(s): 223 - 262 38

Bibliography

[36] D. Krebs (2010) "Electrical Transport and Switching in Phase Change Materials" - Dissertation at RWTH Aachen University, Germany 38, 80, 81

[37] C. Fleck (2010) "Transiente elektronische Effekte waehrend des Threshold Switches in Phasenwechselmaterialien" - Diploma Thesis at RWTH Aachen University, Germany 75, 81, 82

[38] M. Wimmer (2010) "Feldinduzierte Uebergangseffekte in Phasenwechselmaterialien" - Diploma Thesis at RWTH Aachen University, Germany 70, 81, 82

[39] G. Wittenberg, S.J. Wang (2006) "Malleability of Spike-Timing-Dependent Plasticity at the CA3ÐCA1 Synapse" - The Journal of Neuroscience 2006, Volume: 26, Issue: 24, Page(s): 6610 - 6617 42

[40] US Patent Document 6,999,953 by Ovshinsky et al. (2006) 43, 44, 45

[41] E.M. Izhikevich (2006) "Bursting" - Scholarpedia, 1(3):1300., revision 91090. 47, 50, 51, 54, 55

[42] E.M. Izhikevich, N.S. Desai, E.C. Walcott, F.C. Hoppensteadt (2003) "Bursts as a unit of neural information: selective communication via resonance" - Trends in Neuroscience 2003, Volume: 26, Issue: 3, Page(s): 161 - 167 47, 55, 56, 57, 58, 59

[43] J. Lisman (1997) "Bursts as a unit of neural information: making unreliable synapses reliable" - Trends in Neuroscience, Volume: 20, Issue: 1, Page(s): 38 - 43 47, 54, 55

[44] S.E. Gartside, E Hajos-Korcsok, E. Bagdy, L.G. Harsing Jr., T. Sharp, M. Hajos (2000) "Neurochemical and electrophysiological studies on the functional significance of burst firing in serotonergic neurons" Neuroscience 2000, Volume: 98, Issue: 2, Page(s): 295 - 300 47, 54

[45] D.A. Butts, P.O. Kanold, C.J. Shatz (2007) "A burst-based 'Hebbian' learning rule at retinogeniculate synapses links retinal waves to activity-dependent refinement" - PLoS Biology 2007, Volume: 5, Issue: 3, Page(s): 61 47, 55

[46] B. Doiron, M.J. Chacron, L. Maler, A. Longtin, J. Bastian (2003) "Inhibitory feedback required for network oscillatory responses to communication but not prey stimuli" - Nature 2003, Volume: 421, Page(s): 539 - 543 47, 48, 49, 50, 55

[47] N.J. Berman, L. Maler (1999) "Neural architecture of the electrosensory lateral line lobe: adaptations for coincidence detection, a sensory searchlight and frequency-dependent adaptive filtering" - Journal of Experimental Biology 2003, Volume: 202, Page(s): 1243 - 1253 47, 48

Bibliography

[48] L. Maler, E.K. Sas, J. Rogers (1981) "The cytology of the posterior lateral line lobe of high frequency weakly electric fish (Gymnotoidei): differentiation and synaptic specificity in a simple cortex" - Journal of Comparative Neurology 1981, Volume: 195, Issue: 1, Page(s): 87 - 139 47

[49] M.E. Nelson, M.A. MacIver, (1999) "Prey capture in the weakly electric fish Apteronotus leptorhynchus: sensory acquisition strategies and electrosensory consequences" - Journal of Experimental Biology 1999, Volume: 202, Page(s): 1195 - 1203 47

[50] W. Metzner (1999) "Neural circuitry for communication and jamming avoidance in gymnotiform electric fish" - Journal of Experimental Biology 1999, Volume: 202, Page(s): 1365 - 1375 47

[51] P. Dayan, L.F. Abbott (2001) "Theoretical Neuroscience" (MIT Press, Cambridge, Massachusetts). 48, 72

[52] B. Amini, J.W. Clark Jr, C.C. Canavier (1999) "Calcium dynamics underlying pacemaker-like and burst firing oscillation in midbrain dopaminergic neurons: A computational study" - Journal of Neurophysiology 1999, Volume: 82, Page(s): 2249 - 2261 51

[53] X.J. Wang (1999) "Fast burst firing and short-term synaptic plasticity: a model of neocortical chattering neurons" - Neuroscience 1999, Volume: 89, Page(s) :347 - 362 51

[54] J.R. Huguenard, D.A. McCormick (1992) "Simulation of the currents involved in rhythmic oscillation in thalamic relay neurons" - Journal of Neurophysiology, Volume: 68, Page(s): 1373 - 1383 51

[55] R.M. Harris-Warrick, R.E. Flamm R.E. (1987) "Multiple mechanisms of bursting in a conditional bursting neuron" - Journal of Neuroscience 1987, Volume: 7, Page(s): 2113 - 2128 51

[56] E.M. Izhikevich (2007) "Dynamical Systems in Neuroscience: The Geometry of Excitability and Bursting" (The MIT Press, Cambridge, Massachusetts). 51, 52, 53, 59, 60, 61, 62, 63, 72, 73

[57] M.S. Jensen, R. Azouz, Y. Yaari (1994) "Variant firing patterns in rat hippocampal pyramidal cells modulated by extracellular potassium" - Journal of Neurophysiology, Volume: 71, Page(s): 831 - 839 51

[58] H. Su, G. Alroy, E.D. Kirson, Y. Yaari (2001) "Extracellular calcium modulates persistent sodium current-dependent burst-firing in hippocampal pyramidal neurons" - Journal of Neuroscience, Volume: 21, Page(s): 4173 - 4182 51, 54, 72, 75, 76

Bibliography

[59] C. Yue, Y. Yaari (2004) "KCNQ/M channels control spike afterdepolarization and burst generation in hippocampal neurons" - Journal of Neuroscience, Volume: 24, Page(s): 4614 - 4624 51

[60] J.C. Magee, M. Carruth (1999) "Dendritic voltage-gated ion channels regulate the action potential firing mode of hippocampal CA1 pyramidal neurons" - Journal of Neurophysiology, Volume: 82, Page(s): 1895 - 1901 51

[61] R. Azouz, M.S. Jensen, Y. Yaari (1996) "Ionic basis of spike after-depolarization and burst generation in adult rat hippocampal CA1 pyramidal cells" - Journal of Physiology 1996, Volume: 492, Page(s): 211 - 223 51, 75, 76

[62] B. Doiron, C. Laing, A. Longtin, L. Maler (2002) "'Ghostbursting: a novel neuronal burst mechanism" - Journal of Computational Neuroscience 2002, Volume: 12, Page(s): 5 - 25 53

[63] B.W. Connors, M.J. Gutnick (1990) "Intrinsic firing patterns of diverse neocortical neurons" - Trends in Neuroscience 1990, Volume: 13, Page(s): 99 - 104 54

[64] C.M. Gray, D.A. McCormick (1996) "Chattering cells: Superficial pyramidal neurons contributing to the generation of synchronous oscillation in the visual cortex" - Science 1996, Volume: 274, Issue: 5284, Page(s): 109 - 113 54

[65] J.C. Smith, H.H, Ellenberger, K. Ballanyi, D.W. Richter, J.L. Feldman (1991) "Pre-Botzinger complex: a brainstem region that may generate respiratory rhythm in mammals" - Science 1991, Volume: 254, Issue: 5032, Page(s): 726-729 53, 54

[66] W. Kahle, M. Frotscher (2005) "Taschenatlas Anatomie - 3 Nervensystem und Sinnesorgane" - Thieme, ISBN: 3-13-492209-6 53

[67] S.M. Sherman (2001) "Tonic and burst firing: dual modes of thalamocortical relay" - Trends in Neuroscience 2001, Volume: 24, Page(s): 122 - 126 54

[68] F. Gabbiani, W. Metzner, R. Wessel R, Koch C. (1996) "From stimulus encoding to feature extraction in weakly electric fish" - Nature 1996, Volume: 384, Issue: 6609, Page(s): 564 - 567 55

[69] A.M. Oswald, M.J. Chacron, B. Doiron, J. Bastian, L. Maler (2004) "Parallel processing of sensory input by bursts and isolated spikes" - Journal of Neuroscience 2004, Volume: 24, Issue: 18, Page(s): 4351 - 4362 55

[70] N.A. Lesica, G.B. Stanley (2004) "Encoding of natural scene movies by tonic and burst spikes in the lateral geniculate nucleus" - Journal of Neuroscience 2004, Volume: 24, Page(s): 10731 - 10740 55

Bibliography

[71] P. Reinagel, D. Godwin, S.M. Sherman, C. Koch (1999) "Encoding of visual information by LGN bursts" - Journal of Neurophysiology 1999, Volume: 81, Page(s): 2558 - 2569 55

[72] H. Markram, et al. (1998) "Differential signaling via the same axon of neocortical pyramidal neurons" - Proceedings of the National Academy of Sciences of the USA 1998, Volume: 95, Page(s): 5323 - 5328 57

[73] A. Gupta et al. (2000) "Organizing pronciples fpr a diversity of GABAergic interneurons and synapses in the neocortex" - Science 2000, Volume: 287, Page(s): 273 - 278 57

[74] E.S. Fortune, G.J. Rose (2000) "Short-term synaptic plasticity contributes to the temporal filtering of electrosensory information" - Journal of Neuroscience 2000, Volume: 20, Page(s): 7122 - 7130 57

[75] A.M. Thomson (2000) "Molecular frequency filters at central synapses" - Progress in Neurobiology 2000, Volume: 62, Page(s): 159 - 196 57

[76] R.R. Llinas et al. (1991) "*In vitro* neurons in mammalian cortical layer 4 exhibit intrinsic oscillatory activity in the 10 to 50 Hz frequency range" - Proceedings of the National Academy of Sciences of the USA 1991, Volume: 88, Page(s): 897 - 901 57

[77] B. Liu, J.F. Frenzel, (2002) "A CMOS Neuron for VLSI Circuit Implementation of Pulsed Neural Networks" - IECON 2002, Volume: 4, Page(s): 3182 - 3185 65

[78] D. Ielmini, D. Mantegazza, A.L. Lacaita, (2008) "Voltage-Controlled Relaxation oscillation in Phase-Change Memory Devices" - Electron Device Letters, IEEE 2008, Volume: 29, Issue: 6, Page(s): 568 - 570 65, 66, 67, 68, 69, 77, 78, II

[79] M. Anbarasu, M. Wimmer, G. Bruns, M. Salinga, M. Wuttig (2012) "Nanosecond threshold switching of GeTe6 cells and their potential as selector devices" - Applied Physics Letters 2012, Volume: 100, Issue: 14, Page(s): 143505 - 14350-4 68, 71, 75

[80] D. Ielmini, A. L. Lacaita, and D. Mantegazza, (2007) "Recovery and drift dynamics of resistance and threshold voltages in phase-change memories" - IEEE Trans. Electron Devices 2007, Volume: 54, Issue: 2, Page(s): 308 - 315 69

[81] D. Ielmini, D. Mantegazza, A. L. Lacaita, A. Pirovano, and F. Pellizzer (2005) "Parasitic reset in the programming transient of PCMs" - IEEE Electron Device Letters 2005, Volume: 26, Issue: 11, Page(s): 799 - 801 69

[82] D. Ielmini and Y. Zhang, (2007) "Analytical model for subthreshold conduction and threshold switching in chalcogenide-based memory devices" - Journal of Applied Physics 2007, Volume: 102, Issue: 5, Page(s): 054517 - 054517-13 38, 70

Bibliography

[83] K. E. Petersen, D. Adler (1976) "On state of amorphous threshold switches" - Journal of Applied Physics 1976, Volume: 47, Issue: 1, Page(s): 256 263 71

[84] S. Lai, T. Lowrey (2001) "OUM - A 180 nm nonvolatile memory cell element technology for stand alone and embedded applications " - Tech. Dig. - Int. Electron Devices Meet. 2001, Volume: 36, Issue: 5, Page(s): 1 - 4 71

[85] A.L. Hodgkin, B. Katz (1949) "The effect of sodium ions on the electrical activity of the giant axon of the squid" - Journal of Physiology 1949, Volume: 108, Issue: 1, Page(s): 37 - 77 87, 88

[86] A. Pirovano, A.L. Lacaita, A. Benvenuti, F. Pellizzer, R. Bez (2004) "Electronic Switching in Phase-Change Materials" - IEEE Transactions on Electron Devices 2004, Volume: 51, Issue: 3, Page(s): 452 - 459 38, 39

[87] V.G. Karpov, Y.A. Kryukov, S.D. Savransky, I.V. Karpov (2007) "Nucleation switching in phase change memory" - Applied Physics Letters 2007, Volume: 90, Page(s): 123504 - 123504-3 38, 39

[88] S. Lavizzari, D. Sharma, D. Ielmini (2010) "Threshold-Switching Delay Controlled by $1/f$ Current Fluctuations in Phase-Change Memory Devices" - IEEE Transactions on Electron Devices 2010, Volume: 57, Issue: 5, Page(s): 1047 - 1054 84, 85

[89] P. Haenggi (2002) "Stochastic Resonance in Biology - How Noise Can Enhance Detection of Weak Signals and Help Improve Biological Information Processing" - ChemPhysChem 2002, Volume: 3, Page(s): 285 - 290 85

[90] P.E. Greenwood, L.M. Ward, A.D.F. Russel, A. Neiman, F. Moss (2000) "Stochastic Resonance Enhances the Electrosensory Information Available to Paddlefish for Prey Capture" - Physical Review Letters 2000, Volume 84, Issue: 20, Page(s): 4773 - 4776 85

[91] J.E. Levin, J.P. Miller (1996) "Broadband neural encoding in the cricket cercal sensory system enhanced by stochastic resonance" - Nature 1996, Volume: 380, Page(s): 165 - 168 85

[92] J.A. Freund, L. Schimansky-Geier, B. Beisner, A. Neiman, D.F. Russel, T. Yakusheva, F. Moss (2002) "Behavioral Stochastic Resonance: How the Noise from a *Daphnia* Swarm Enhances Individual Prey Capture by Juvenile Paddlefish" - Journal of Theoretical Biology (2002), Volume: 214, Page(s): 71 - 83 85

[93] J.K. Douglas, L. Wilkens, E. Pantazelou, F. Moss (1993) "Noise enhancement of information transfer in crayfish mechanoreceptors by stochastic resonance" - Nature 1993, Volume: 365, Page(s): 337 - 340 85

Bibliography

[94] D.F. Russel, L.A. Wilkens, F. Moss (1999) "Use of behavioural stochastic resonance by paddle fish for feeding" - Nature 1999, Volume: 402, Page(s): 291 - 294 85

[95] E. Simonotto, M. Riani, C. Seife, M. Roberts, J. Twitty, F. Moss (1997) "Visual Perception of Stochastic Resonance" - Physical Review Letters 1997, Volume: 78, Issue: 6, Page(s): 1186 - 1189 85

[96] K. Wiesenfeld, F. Moss (1995) "Stochastic resonance and the benefits of noise: from ice ages to crayfish and SQUIDs" - Nature 1995, Volume: 373, Page(s): 33 - 36 85

[97] M. Juusola, H.P.C. Robinson, G.G. de Polavieja (2007) "Coding with spike shapes and graded potentials in cortical networks - BioEssays 2007, Volume: 29, Issue: 2, Page(s): 178 - 187 72

Acknowledgement

At this point, the author would like to express his gratitude to a number of people who contributed to this master's thesis project in various ways.

First and foremost, I would like to thank Prof. Dr. Matthias Wuttig for the possibility to write this master's thesis at the I. Physikalisches Institut (IA). His short, yet inspirational talk about the human brain as a storage medium, which he incidentally gave in one of his lectures about solid state physics, brought this field of research to my attention for the very first time and I am highly fascinated with it ever since. I would also like to thank Dr. Martin Salinga for the possibility to write this master's thesis in his workgroup and for deepening my understanding of scientific work in a team. Special thanks goes to my direct advisor Dipl. Phys. Martin Wimmer who always had an open ear which he lent to me for several fruitful discussions. In addition, I would like to thank my bureau colleagues Christoph Schmitz, Sascha Cramer, Christian Dellen, Simon Ritz, Fabian Wendt, Lukas Kuepper and Carl Henning Lubba for quite some funny moments in the bureau and many exciting games of table soccer during lunchtime. I would like to thank Carl Henning Lubba in particular for the pleasant collaboration during his UROP project in the workgroup.

Furthermore, I would like to thank my girlfriend Anna Schwingenheuer for all her support during some difficult times in the past year.

Zu guter Letzt möchte ich an dieser Stelle die Gelegenheit nutzen und den Versuch unternehmen sowohl meiner aufrichtigen Dankbarkeit als auch meiner tiefsten Anerkennung gebührend Ausdruck zu verleihen, für die Leistung zweier Menschen in meinem Leben, ohne die es mir niemals möglich gewesen wäre heute diese Zeilen schreiben zu können. Ich danke Euch für Euer großes Verständnis, für Eure aufopferungsvolle Hingabe und für Eure unerschöpfliche Kraft, mit der Ihr mich mein ganzes, bisheriges Leben lang unterstützt habt, ganz besonders aber danke ich Euch für die Gewissheit Euer immer geliebter Sohn zu sein.

    Ich danke Euch, meine lieben Eltern!